FOWL PLAY

Sally Coulthard is a bestselling author of books about natural history and rural life including *The Barn*, *A Short History of the World According to Sheep*, *The Book of the Barn Owl*, *The Hedgehog Handbook* and over twenty more titles. She lives on a Yorkshire smallholding which she shares with her husband, three girls and an assortment of unruly animals.

Also by Sally Coulthard

The Barn

A Short History of the World According to Sheep

The Book of the Barn Owl

The Book of the Earthworm

The Bee Bible

The Hedgehog Handbook

The Little Book of Snow

The Little Book of Building Fires

FOWL PLAY

A History *of the* Chicken *from* Dinosaur *to* Dinner Plate

SALLY COULTHARD

An Apollo Book

First published in the UK in 2022 by Head of Zeus
This paperback edition first published in 2023 by Head of Zeus,
part of Bloomsbury Publishing Plc

9 7 5 3 1 2 4 6 8

A catalogue record for this book is available from the British Library.

ISBN (PB): 9781801104487
ISBN (E): 9781801104494

Typeset by Ed Pickford
Illustrations by © Alice Pattullo

Printed and bound in Great Britain by
CPI Group (UK) Ltd, Croydon CRO 4YY

Head of Zeus
5–8 Hardwick Street
London EC1R 4RG

WWW.HEADOFZEUS.COM

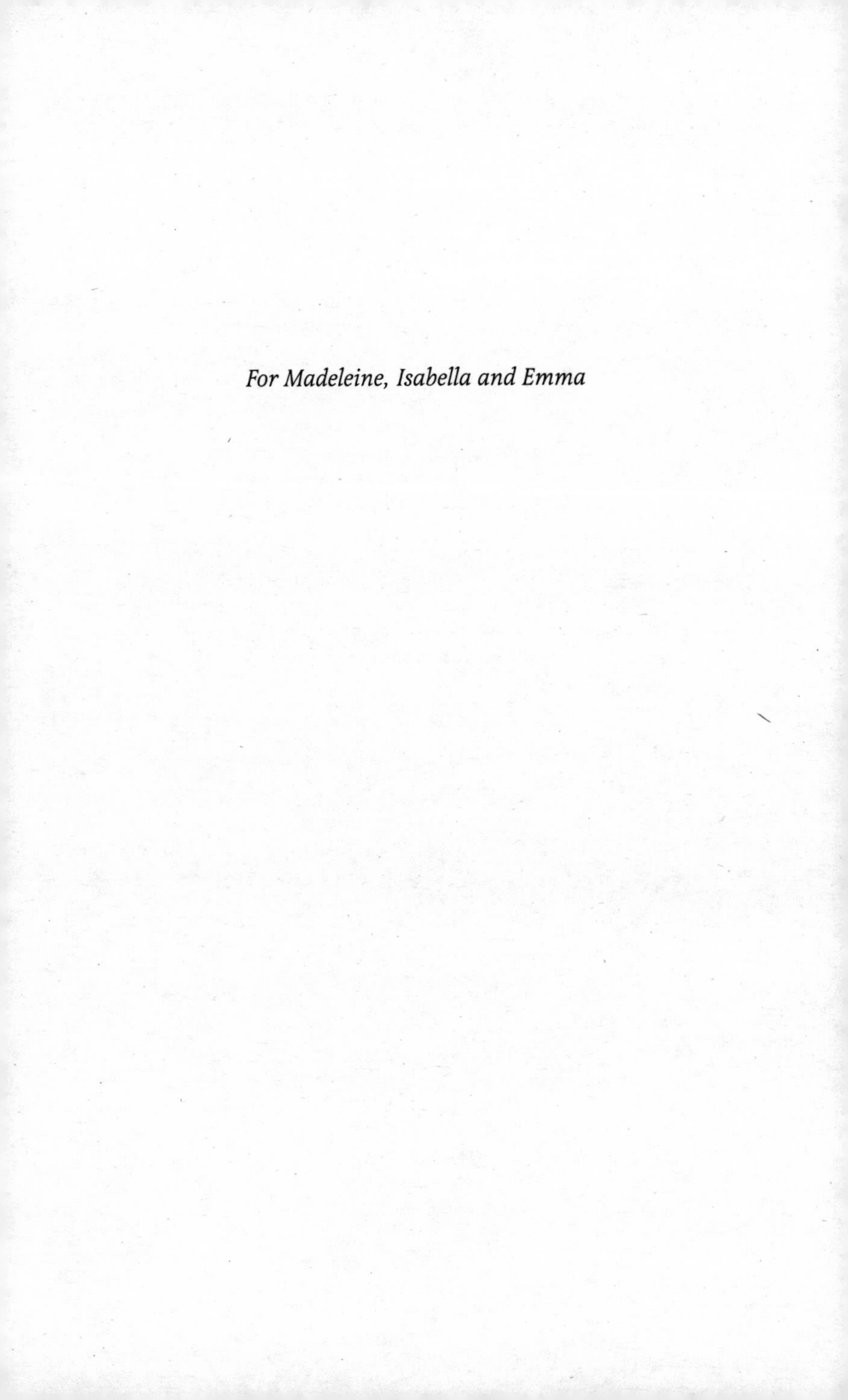

For Madeleine, Isabella and Emma

Wondrous the Gods, more wondrous are the Men,
More Wondrous, Wondrous still the Cock and Hen.

William Blake (1757-1827)

CONTENTS

Lavender Pekin

INTRODUCTION

It's raining and the hens look particularly peeved. I often think their feathers are poorly suited to the English weather. They sit, huddled in the entrance to the hay barn, like disgruntled pensioners waiting for a bus. Winter must be disorientating for a creature descended from the balmy jungles of Southeast Asia. Still, Andy, our formidable cockerel, keeps up their spirits with his endless provocative dances – a jaunty sailor's jig with a sideways hop and one wing raised. Few of them can resist his conscientious 'tidbitting', making gentle clucks to his ladies to alert them to a newly discovered morsel.

The cockerel was named by my youngest daughter after my father, Andrew, a man of immense character and kindness. We raised Andy from an egg, with the intention of hand-rearing a male bird as an interesting project. Naming livestock is fatal, however, and almost always ends in heartbreak as nature picks them off, one by one, as a lesson against favouritism. Against all odds, Andy has proved indomitable and unusually tame. The farm is now graced with a bird who defends his flock with kamikaze bravery but feels no shame in coming to the window ledge for a tickle under his wattle. I know of no other cockerel who will be cuddled tightly and maternally jogged up and down, like a pudgy baby.

I've kept dozens of chickens over the years. Each had its own unique character and, as with humans, some were more memorable and likeable than others. One fat hen, a black Marans called Brenda, seemed to prefer the children to her flock mates and would spend hours keeping the girls company, only wandering off to lay the occasional chocolate-brown egg. Corralled into kids' pretend play, Brenda would endure endless tea parties and tepee adventures. I have an abiding memory of Brenda hurtling down a snowbound field in a plastic sledge, only to come to a gentle halt, step out and carry on pecking as if nothing was amiss.

We also once inherited ten pure-white Dorking hens from a redoubtable but kindly dowager of a very large country house. Like Mrs Pumphrey's Tricki Woo, the birds had known nothing but luxury and laid perfectly white eggs only when the choicest morsels were provided and the mood struck. Their pampered early life, however, put them in good stead. The whole flock lived an astonishingly long life. All made it to ten and the eldest reached thirteen years of age before finally falling off her perch, both literally and metaphorically.

And then there was Cato, a Rhode Island Red of startling intelligence who excelled at finding new and unusual ways to sneak into the farmhouse. She was the mistress of stealth, tiptoeing through open windows and doors left ajar, avoiding detection while helping herself to the dog bowl. She was also fantastic at hiding. Open a cupboard or shed and she'd leap out. Cato even made it down the farm track hidden in the footwell of a DPD van. She got half a mile away before a very surprised courier caught a glimpse of red feathers out of the corner of his eye and was forced to turn around.

I believe my flock and, indeed, all chickens are truly fascinating. Of all animals, they perhaps best represent the strange and often contradictory way we humans treat other species. They're both beloved pet and cheap commodity, symbol of rural simplicity and icon for 'misery meat' and the industrialisation of food. The chicken is also a bird most of us are both deeply familiar with and yet know very little about. Its evolutionary past is full of surprises, as is the bird's journey from jungle to domestication.

From fighter to farmed food, scientific tool to pampered pet, the chicken has been forced to bend to almost every human whim. Great civilisations valued the chicken for its many uses, from ceremonial rites to celebratory feasts, cruel pastimes to status symbols. In its various incarnations - egg, chick, pullet, hen, cockerel - the chicken has proved a shorthand for human relationships and emotions. We use chicken-related words to express a whole range of feelings and situations - from mother hens and feeling broody to cockiness and flock mentality. Religion and superstition have also dragged the chicken centre stage, whether the bird likes it or not. A vast array of belief systems have chickens, eggs and cockerels at their heart, handy metaphors for rebirth and innocence but also vigilance and protection.

There are, at any one time, over twenty billion chickens living on Earth – that's three chickens for every person alive. And, for a creature that doesn't fly very far, it's somehow managed to colonise the world. Only one continent, Antarctica, is fowl free. From the frozen wastes of Siberia to the Falkland Islands in the middle of the Atlantic Ocean, you'll find chickens scratching around in the dirt. This humble bird's voyage around the world is inextricably linked to human exploration, trade, diet and exploitation. At every stage, we and chickens made that remarkable journey together.

This is that story.

I

SURVIVORS

I Am Chicken, Hear Me Roar

Red Junglefowl

On a day like any other, sixty-six million years ago, the world ground to a halt. An asteroid the size of a city scorched through the atmosphere at more than forty times the speed of sound and slammed into Earth. When it hit, just off the coast of Mexico, the resulting explosion was seven billion times more powerful than Hiroshima and ripped a 150-kilometre-wide hole in the planet's crust.

The collision – now known as the Chicxulub impactor – sent immediate shockwaves across the world's surface. Earthquakes and volcanic eruptions tore the land apart, wildfires flattened forests and giant tsunamis engulfed the coastlines. This violent upheaval, however, was nothing compared with what was to follow. The explosion also injected billions of tons of debris and poisonous gases, including sulphur and carbon dioxide, into the air, choking the atmosphere and sending the climate into freefall. Three-quarters of all life on Earth died. The reign of the dinosaurs, which had ruled the planet for over 170 million years, came to an abrupt and catastrophic end.

Well, nearly. One form of dinosaur miraculously survived this mass extinction. Those dinosaurs were the ancestors of all the birds that currently flap, peck and waddle across the world. What's more, if you want to see the creature that most closely resembles the nightmarish, thundering beast that was the *Tyrannosaurus rex*, you'll find it absent-mindedly scratching around in a farmyard. Chickens are the dinosaurs that didn't die.

In 1861 a fossil was discovered in a German quarry that sent the scientific community into a spin. The magpie-sized specimen looked like a creature from Greek mythology – half bird, half reptile – a sensational hybrid of feathers, wings, claws, teeth and a bony tail. Only a few years earlier, Charles Darwin had published *On the Origin of Species* and predicted the existence of 'missing links', fossils that showed the evolution of one species into another. Here, in front of the whole world, was an ancient creature that seemed to prove his point. With its blend of reptilian and avian features, here was proof of the transition between dinosaurs and birds. The fossilised animal was named *Archaeopteryx*, meaning 'ancient wing'. And, at 147 million years old, *Archaeopteryx* was hailed as marking the beginning of the 'Age of Birds' – a key moment when dinosaurs finally took to the sky.

Birds are thought to have evolved from theropods, a family of dinosaurs that also included the thunderous T.Rex and sickle-clawed *Velociraptor*. Despite their infamy as flesh-tearing carnivores, theropods also had many bird-like qualities that had been slowly evolving since the family's emergence over 200 million years ago, well before the arrival of *Archaeopteryx*. These avian-like characteristics included laying eggs, light and hollow bones, hinged ankles, walking on two feet and, in some cases, even feathers. One species of *Tyrannosaurus*, the *Yutyrannus*, seems to have been covered in plumage from head to foot, while fossils show that *Velociraptors* also had feathered forearms or 'proto-wings'. Many palaeontologists now believe that the appearance of feathers on dinosaurs wasn't initially anything to do with flight and more likely developed as a way to retain body heat, rather like hair. These downy-like feathers evolved over time into wing-like structures but again, probably not for flight but for display or intimidation. Most of the early feathered dinosaurs would have simply been too big to fly.

Only when a certain group of theropods, including *Archaeopteryx*, evolved the winning combination of feathered wings and a small body size around 150 million years ago could flight, or at least gliding, become a reality. And while *Archaeopteryx* is still viewed by most palaeontologists as the earliest bird, it seems that in the eighty-million-year interval between the *Archaeopteryx* and the asteroid impact, numerous other prehistoric birds evolved and lived side by side with dinosaurs. In the last two decades alone, more than three hundred new fossil species of birds have been named, many carrying bits of evolutionary baggage – such as tiny teeth or claws on their wings – that show the slow shift from reptile to bird. The fact that scientists still have a hard time distinguishing between the earliest true birds and many avian-like dinosaurs shows how complex and gradual the transition must have been.

When the asteroid hit Earth, its devastation was indiscriminate. The impact and its fallout wiped out the majority of all plant and animal life. This included most of these nascent bird species but, crucially, not all. In the post-apocalyptic landscape, four distinct lineages emerged and went on to become the ancestors of all birds today: Anseriformes (which now include waterfowl such as ducks, geese and swans); Palaeognathae (birds that excel at running rather than flying, such as emus and ostriches); Galliformes (land fowl such as chickens and pheasants); and Neoaves (basically everything else, from owls to hummingbirds).

No one knows why certain birds survived the impact when most didn't, but three ideas have been proposed. One theory is that body size allowed particular species to thrive – all the prehistoric birds who made it through the asteroid impact were no bigger than ducks. Having a small body size helped in two ways – not only do small creatures need less food to survive, which is critical in a decimated landscape, but little animals tend to breed faster, helping populations quickly recover.[1]

A second hypothesis is that only ground-dwelling birds managed to survive in the post-asteroid landscape, while all those creatures who relied on forests for food and shelter went extinct. A handful of birds may have managed to eke out a living clawing around in the dirt or along the coastlines for morsels. Based on findings from the pollen record, fossils and modern bird ecology, the science suggests that just a few ground-dwelling species survived the impact but then evolved to reoccupy all the different ecological niches that blossomed out of the devastation.[2]

The recent discovery of 'Wonderchicken', a fossilised bird alive just before the mass extinction, seems to confirm this theory. The bird was found in a quarry on the Netherlands-Belgium border - a region that would have once been covered with tropical beaches and shallow seas - and analysed by scientists at the University of Cambridge. From its remains, 'Wonderchicken' seems to have possessed long, sandpiper-like legs, suited to life as a shore dweller, but also a curious mix of both chicken and duck features on its skull. The discovery is fascinating for so many reasons, not least because 'Wonderchicken' may have been a common ancestor of both Galliformes and Anseriformes, before chickens and ducks went their separate evolutionary ways. Its small body size at around 400 grams, and preference for a life by the sea, not in the tree, may also have protected it from the fate of many of its feathered contemporaries.[3]

A third possibility is that some birds, even before the asteroid, had started to develop a new, game-changing facial feature. At least twenty million years before the mass extinction, some birds had started to lose their dinosaurian teeth and instead develop beaks.[4] Beaks allowed ancient birds to eat a much more varied diet, including fruit, insects and seeds. In the aftermath of the disaster, this generalist diet became even more important, especially when food sources were so scarce. Some scientists believe that the ability to feed on the tough seeds and nuts left behind in the otherwise denuded forests gave beaked birds enough nutrition to survive until the vegetation slowly returned.

In all likelihood, it was a combination of all these winning characteristics that allowed certain species of birds to survive when so many perished. The ability to eat a wide range of foods, live away from forests, and survive on a modest calorific intake – all these factors and others could have enabled a handful of birds to shift and flex with the ever-changing ecosystem around them. For the few birds who made it through the asteroid and its fallout, there was a brave new world to conquer.

So how does the chicken fit into this picture? Back in 2008, *Science* published an astonishing finding. Researchers had managed to uncover tiny fragments of unfossilised material inside a *Tyrannosaurus rex* bone. The laboratory couldn't retrieve any ancient DNA from the sample but its scientists did manage to harvest collagen protein molecules. When these were compared to samples from twenty-one living animals, including humans, chimps, alligators and salmon, the scientists could create a family tree based on protein sequences. Those creatures with similar sequences must be closely related, surmised the scientists, while differences in the sequences indicated which animals had diverged a long time ago. T.Rex's molecules seemed to confirm what palaeontologists and fossil hunters had long suspected – that birds were descended from meat-eating theropods. What came as more of a surprise was that the protein sequence from the most iconic and feared dinosaur of all time was closest to that of a modern chicken.[5]

Despite the excitement generated by the study, the results were highly controversial. Accusations of contaminated material and poor science drowned the initial findings, but in 2014 a different study again suggested a close link between chickens and dinosaurs. Research by the University of Kent looked at the chromosomes of a number of different modern birds – including the chicken, turkey, Pekin duck, zebra finch and budgerigar – and discovered that the chromosomes of chickens and ostriches had undergone the least number of changes from their dinosaur beginnings.[6] The birds who survived the mass extinction event are thought to have undergone a rapid burst of evolution in the years directly after. The survivors hit the jackpot in evolutionary terms, colonising new ecological niches and rapidly mutating and evolving into the dazzlingly biodiverse ten thousand avian species we have today. Not all birds changed as much as others, however. Out of all the birds in the study, the chicken remained one of the two most genetically similar to its dinosaur ancestors, despite thousands of years of domestication and interbreeding.

Chickens have proved to be a handy candidate for the study of the evolution of birds. Not only do they share a common heritage with dinosaurs but, perhaps to their detriment, they're also a much more pliable experimental subject than a 100-kilogram ostrich. As we'll see later in the book, poultry science and high-tech farming is big business and in 2004 the chicken, with its enormous commercial potential, was the first bird to have its genome sequenced. This combination of commonality with dinosaurs and commercial availability has led scientists to try to model the chicken's long transition from theropod to backyard bird. More controversially, by tweaking chicken development, both in the embryo and in live animals, scientists seem to be able to 'switch on' long-lost dinosaurian traits in the humble hen.

Take locomotion. Fossils can only tell you so much. For years, scientists have pondered over how bipedal dinosaurs – such as the T.Rex – would have moved. In a bid to solve the mystery, researchers from the universities of Chile and Chicago strapped fake dinosaur tails to chickens and recorded the results.[7] Looking not unlike a toilet plunger, each 'tail' – which was made from a wooden stick – was Velcroed to a bird's bottom as it grew from a young chick to an adult. Every five days, the sticks were replaced with slightly bigger ones, to mimic the growing tail of a theropod.

Ordinarily, chickens walk in a crouched position, flexing from the knees. As they walk, chickens hold their femora (the big bone at the top of each thigh) almost horizontal to the ground and most of the leg movement is concentrated at the knee joint. The study revealed that when a chicken carries an extra weight on its rear, its centre of balance also shifts and the bird walks differently. To compensate for the weight of the tail, each chicken walked not from the knee but rather by swinging the entire leg from the hip, not unlike a cowboy. This hip-driven locomotion also made the chickens straighten their legs as they moved. The gait of the T.Rex was revealed.

So why did chickens and other birds lose their dinosaurian tails? The short answer is that a meaty tail makes flying tricky. Over time, some theropods' tails evolved into a stump and the last few tail vertebrae fused together to form something called a *pygostyle*. On roast chicken, the fatty tissue around this bit is often known as the 'parson's nose'. Amazingly, however, the chicken has retained its genetic instructions for growing a long dinosaur tail. Back in 2007, Hans Larsson, a palaeontologist from McGill University in Canada, discovered that two-day-old chick embryos had sixteen vertebrae, nine more than when the fully formed chick hatches. In other words, the chicken embryos still have dinosaur-like tails in the early days of growth but reabsorb most of the vertebrae during incubation. Larsson then realised that if he could turn off the genetic signal that triggered the tail to be reabsorbed, the chick would carry on growing a reptilian-like tail, turning back the clock on millions of years of bird evolution. And while a chick has yet to be born with a full-length lizardy, whip-like tail, Larsson was able to extend the tail by an entire three vertebrae.

If chickens retained the genetic instructions for a dinosaur tail, the next logical question for researchers was whether they could recreate other features of a dinosaur. The feathered ancestors to modern birds, such as *Archaeopteryx,* didn't have beaks, but rather snouts. To understand how one mouth shape changed into another, a group of researchers in the US began manipulating the facial structure of chicken embryos, switching off the unique cluster of genes that tells the face to grow a beak. In the process, they managed to create a chicken embryo with a dinosaur-like snout, not unlike a small *Velociraptor*.[8] In a similar study, researchers also found they could get chicken embryos to start sprouting tiny teeth. The latent genes were there and, when 'turned on', resulted in the beginnings of cone-like, sabre-shaped teeth so familiar in dinosaur fossils.[9] Both studies showed that modern birds might have lost their snout and teeth over time but not the ability to grow either.

Over the past few years, this kind of 'atavism activation' – the reawakening of dormant traits – has attracted significant public interest. The idea that, with the help of the humble hen, science will be able to recreate a dinosaur seems like the plot from *Jurassic Park*. And yet tinkering with inactive sections of the chicken's own genetic code is perhaps the best chance scientists have of recreating the past. Dinosaur DNA breaks down too quickly for scientists to extract it from a fossil. Therefore, building a dinosaur by tweaking a chicken may be the most scientifically plausible option. Alongside ancestral characteristics such as teeth, snouts and tails, researchers are also delving into the genetic mechanisms that turn chicken wings back into theropod forearms (by switching off the genetic instructions that fuse a dinosaur's three 'fingers' into a modern wing), scales into feathers, and chickens' limbs into theropod-like toes and legs. None of the experiments so far have attempted to, or even wanted to, produce a living creature from a modified chicken embryo but if it happens, the chicken will have a starring role. The age of the chickenosaurus may yet come.

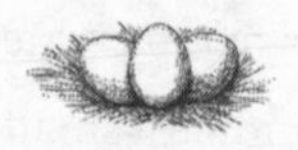

One of the more curious and eyebrow-raising scientific questions for palaeontologists who study avian evolution is how the cockerel lost his penis. Of the birds that survived the asteroid, Palaeognathae (the ostriches and emus) and the Anseriformes (the waterfowl) all have remarkably pronounced phalluses. Indeed, most ducks are so generously endowed that their corkscrewed penises measure half the length of their bodies. The enviably well-hung Argentine lake duck (*Oxyura vittata*) has a penis the same length as its body, at nearly half a metre long. It's one of nature's delicious ironies that the male chicken, whose name is used as slang for the male anatomy, doesn't actually have a penis.

The chicken is not alone, however, in its emasculation. Over 90 per cent of all modern male birds have little or nothing in terms of an appendage, despite all of them needing to reproduce by internal fertilisation – in other words, getting their sperm inside the body of the female. Both sexes, instead, have to rely on simple genital openings called 'cloacae', which they press together during reproduction. Muscular contractions pump the sperm into the female in a move called a 'cloacal kiss'. The cloaca is the same hole that chickens use for excretion and laying eggs. Researchers led by Martin Cohn, a biologist at the University of Florida, recently discovered that something was going on inside the chicken egg that disrupted the growth of the penis. For the first eight days of growth, the chick embryo grows a penis similar to a duck, but on the ninth day the process stops and its nascent genitals begin to shrink. The trigger for this event is an increase in levels of a protein called Bmp4, which seems to promote cell death in the penis. So why did the chicken and so many other birds lose their tackle, while some – such as ducks and emus – keep theirs?

There are two current avenues of thought. Male birds with penises, such as ducks, often use violent force when it comes to mating. Anyone who's watched drakes and ducks will know that the males can be so vigorous in their attempts to have sex with a female that she can be stressed or even injured in the process. 'Sexual coercion' as a mating strategy might benefit the male of the species – as he will maximise the number of offspring he has – but female birds often prefer to be more selective about who they mate with, picking the best partner in terms of size, health or temperament. Among ducks and geese, this war of the sexes has resulted in a kind of 'arms race' between males and females, with females developing several complex strategies to avoid copulation including evolving different vaginal shapes to thwart the unwanted attention of males with long, elaborate phalluses. For the chicken (and many other birds), this preference for a method of reproduction that requires cooperation, not brute force, may have resulted in females deliberately selecting less amply endowed males. Over time, the cockerel lost his need for any penis at all.[10]

Cohn, however, has a different idea: that the disappearance of the cockerel's penis may have just been collateral damage in the transformation from dinosaur to modern bird. The phallus may have been lost as part of a curious devil's bargain, a swap for other evolutionary developments such as a conical beak shape, or loss of teeth, all of which are also affected by Bmp proteins.[11] The cockerel's loss, however, was undoubtedly the hen's gain. The courtship ritual of the chicken is now, in most cases, a gentler affair and a far cry from the bullying coercion of either its modern cousins – waterfowl and flightless birds – or their common ancestors.

The cockerel must now win the attention of the hen by putting on an alluring performance. Typically, he will initiate mating with a special display, spreading one wing and doing a little circular dance. If the hen likes what she sees, she'll lower her body to the ground ready for the cockerel to jump on. If his dance moves don't have the desired effect, he might quickly switch to phase two, giving a feeding call and pecking at the ground – a behaviour called tidbitting*. If either of these methods is successful, the cockerel will grab the hen's neck feathers in his beak and balance on her back, dipping his tail so their cloacae meet. A few ruffled feathers later, the whole thing is over and both parties get back to scratching the ground.

Interestingly, like many other evolutionary traits that have been 'switched off' over time, the aggressive sexual behaviour of the cockerel's ancient ancestor may still lurk in his genetic code. Studies in broiler chickens – birds bred specifically for meat (see Chapter 7) – have revealed that some cockerels are beginning to lose their seductive dance moves and, even worse,

* Tidbit or titbit? One of the first uses of the word comes from *A Description of the Hundred of Berkeley in the County of Gloucester and of Its Inhabitants* (1639) by the antiquarian John Smyth. 'A tyd bit, i.e. a speciall morsell reserved to eat at last.' The British changed the spelling to 'titbit' over time but the Americans still use the original, correct version – tidbit. 'Tyd' or 'tid' meant tender or fond, 'bit' referred to a morsel or small piece.

reverting to forced sex. With the courtship ritual gone, broiler hens have to endure violent attacks from their mates and those that flee are being maimed or clawed to death by cockerels who seem to have become super-aggressive. When asked to look into the phenomenon, Ian Duncan, Professor Emeritus of Animal Welfare at the University of Guelph, Ontario, suggested that the kind of intensive genetic breeding that takes place in industrial-scale poultry farming – for traits such as generous breast meat or quick growth – may be inadvertently tampering with the genes that affect mating behaviour.[12] The meatier the male bird, it seems, the less gallant its behaviour.

The chicken helps us look deeply into the past, to find common ground among dinosaurs and birds, but it's also interesting to think about the ways that the chicken broke ranks and became the distinctive bird that we know today. As we've already learned, chickens are part of an order of birds called Galliformes. It's a large and fascinating group, containing species as varied as turkeys, ptarmigans and peacocks, but all share a raft of similar traits. The majority of Galliformes have plump bodies with relatively short wings. They're adapted to a life scratching around on the ground, rather than soaring the skies, with stout legs and strong four-toed feet (although some chicken breeds have five toes). Galliformes can fly to varying degrees but rarely migrate or flap for long distances. Most rely on walking to get about but will perform rapid, near-vertical bursts of short flight to make an especially quick escape or reach a night-time perch – the kind of flap-flap-flap-glide flying a panicked pheasant does when it breaks cover. Being airborne is often thought of as the defining characteristic of being a bird but many avian species, including chickens, aren't skilled flyers. Some, like penguins, can't fly at all.

If the entire order of Galliformes were represented as a tree, the largest branch would be Phasianidae, a group of birds that reads like the contents of an Englishman's game larder and includes quail, pheasant, grouse and partridge. In this group, however, are also wild junglefowl, *Phasianus gallus,* a deeply exotic, dazzling family of birds that thrives in the balmy forests of Southeast Asia, India, Sri Lanka and Bangladesh. Consisting of four species – the grey (*Gallus sonneratii),* the green (*Gallus varius*), the Sri Lankan (*Gallus lafayettii*) and the red junglefowl (*Gallus gallus*) – this remarkable family resembles a cross between a handsome chicken and a shy pheasant. The males are highly colourful, almost rainbow-like, with slender bodies, an explosion of shiny, petrol-dark tail feathers and bright-red combs. And, as with pheasants, the females are smaller and duller in colour than their mates, the perfect camouflage for a secretive ground-nesting bird.

Before the advent of modern DNA analysis, one of the challenges that faced early naturalists was trying to establish the true ancestor of domestic chickens. The physical similarities between Asian junglefowl and backyard poultry were undeniable, but it wasn't clear if they were closely related. The huge diversity of plumage, colour, body size, tail formation and other morphological traits among domestic breeds was also confusing. Did the chicken have one true ancestor or did all this variation come from multiple ancestors and different regions? Writing in the late eighteenth century, one of the greatest naturalists of the Enlightenment, Georges-Louis Leclerc, wrestled with this very problem. Not only was the mystery of the 'domestic cock' one of scientific interest but it was starting to feel like a matter of professional pride: 'This bird,' he wrote in his *Natural History of Birds*, 'though a domestic and the most common bird of all, is still, perhaps, not sufficiently known [...] But if the cock be too little known by the bulk of men, what embarrassment must it give to a methodical naturalist who is never satisfied till he refer every object to his classes and genera?'[13]

Leclerc was baffled by the rich assortment of 'foreign breeds' that had been brought to his attention by collectors, and colourful accounts from travellers. Specimens of chickens with almost mythic-sounding names, such as 'The Dwarf Hen of Java', the 'Frizzled Cock' or the 'The Silky Hen of Japan', displayed such a myriad of features he was finding it difficult to imagine a common ancestor for them all. 'Amidst the immense number of breeds of the gallinaceous tribe how shall we determine the original stock?' he wrote, with not a little dismay. 'So many circumstances have operated, so many accidents have occurred, the attention and even the whim of man have so much multiplied the varieties that it seems extremely difficult to trace them to their source.'[14]

After some deliberation, Leclerc plumped for what biologists call a 'polyphyletic' origin. Domestic chickens, he concluded, must have come from multiple ancestors; one single species could surely not have spawned both the tiny pigeon-sized 'Little English Hen' and the 10-pound 'Cock of Padua'. In 1813, Dutch zoologist Coenraad Jacob Temminck came to the same conclusion, suggesting that all domestic poultry breeds had descended from six different wild species, a view shared two decades later by eminent natural scientists Georges Cuvier and René Primevère Lesson.

Charles Darwin is often credited as the first scientist to reject the 'polyphyletic' theory and suggest that all chickens, instead, descended from just one species – the red junglefowl. That honour, however, goes instead to Edward Blyth, a less well-known English zoologist who spent most of his adult life in India as the poorly paid curator of the Museum of the Royal Asiatic Society of Bengal. Blyth was particularly fascinated with exotic birds, a passion that later earned him the title 'Father of Indian Ornithology', but his interest was also piqued by the origins of the domesticated chicken.

In 1847, in a critique of a fellow academic's paper on the 'Birds of Calcutta', Blyth confidently announced that the 'Bengal jungle fowl' – a then common name for the red junglefowl – was 'beyond question, the exclusive aboriginal stock from which the whole of our domestic varieties of common poultry have descended'. It was a breathtakingly bold claim, not least because Blyth had nothing to go on but observation. 'However different these may be, whether the silky fowl of China, the gigantic Chittagong race, or the feather-legged bantams of Burmah, &c., their *voice* at once and unmistakably proclaims their origin [...]

besides that we continually meet with common domestic cocks which correspond, feather by feather, with the wild bird.'[15] All the differences in domestic chickens could be attributed, Blyth believed, to human interference or, as Leclerc had called it, 'the whim of man'. 'You may rest perfectly assured', Blyth wrote four years later, 'that there is no wild silky fowl, or feather-legged, or crested, or black-skinned, or gigantic, frizzled'.[16]

In the following years, Darwin and Blyth wrote affectionately to each other. In 1855, Blyth repeated his assertion in a letter to his learned friend that the red junglefowl 'essentially conforms to the type of the domestic fowl in all its multitudinous varieties, fully as much so as the Mallard does to the domestic drake; or the wild to the tame Turkey!'[17] When Darwin published *On the Origin of Species* in 1859, he gave full credit to Blyth for the idea that all 'breeds of poultry have proceeded from the common wild Indian fowl'. But almost as soon as the book had gone to press, Darwin did a spectacular U-turn and began to doubt his own words. Darwin was particularly troubled by the existence of 'gigantic chickens', breeds of huge fowl that had been described by merchant traders since the early seventeenth century, dead specimens of which he had also examined in the British Museum. At more than twice the height of the average domestic chicken, with strong muscular legs and long necks, the existence of '*Gallus giganteus*' – a bird said to be so tall it could peck grain from the top of a beer barrel – made Darwin wonder whether there might be more ancestors, perhaps undiscovered or extinct, that could upset Blyth's theory.

Thankfully, by 1868 and the publication of *The Variation of Animals and Plants under Domestication*, Darwin had made up his mind. With a rigorous combination of comparative analysis of bird skeletons, cross-breeding of his own chickens, and not a little help from his friend, fellow naturalist and expert poultry breeder William Tegetmeier, Darwin concluded that all domestic chickens, even the whoppers, 'seem all to have diverged by independent and different roads from a single type'.[18] Darwin was also convinced that the red junglefowl was the sole ancestor because crosses between it and domestic chickens produced healthy offspring, while chicks created from domestic poultry mated with other species of junglefowl were almost always infertile. 'From the extremely close resemblance', he wrote, 'in colour, general structure, and especially in voice [...]; from their fertility, as far as this has been ascertained [...] we may confidently look at it as the parent of the most typical of all the domestic breeds.'[19]

Despite Darwin's conclusions, the debate continued. Naturalists flip-flopped between single and multiple ancestral origins for the chicken throughout the late nineteenth and twentieth centuries. Indeed, it wasn't until the 1990s, and the availability of genetic analysis, that a definitive answer could finally be revealed. By looking into the DNA of both chickens and wild junglefowl, scientists assessed their relatedness and determined that – categorically – the red junglefowl was the main ancestral species of all modern chicken breeds.[20] As with all family trees, however, with every generation things get more complicated. More recent studies have shown that although the red junglefowl is the 'grandmother' of the modern chicken, other species of junglefowl have also contributed their genetic material to modern domestic poultry through later breeding between wild and domestic populations.

Some chickens, for example, that are now kept for both egg and meat production in the Western world have yellow skin under their feathers and yellow legs. Curiously, red junglefowl only have the genetic code for white skin. When researchers searched for the location of the yellow-skin gene, they found it in a surprising place – in the DNA of grey junglefowl. Red and grey junglefowl are separate species and don't mate in the wild, leading researchers to conclude that grey junglefowl could have only contributed their 'yellow-skinned' genetic material after humans had started domesticating chickens, bringing different species of junglefowl, and their hybrid offspring, into contact with each other.[21]

Although the debate concerning the genetic ancestor of the chicken was settled, the place and time of the bird's domestication remained to be discovered. One reason for this was that, prior to reliable genetic testing, trying to distinguish between the archaeological remains of a junglefowl and those of a modern chicken was extremely difficult. Red junglefowl (*Gallus gallus*) and chickens (*Gallus gallus domesticus*) are actually very similar. The few things that separate the two creatures are, more often than not, almost impossible to detect in a pile of dry, dusty ancient bones. The standard scientific criteria for wild junglefowl in comparison to the domestic chicken include horizontal tail feathers (as opposed to a perky, upright tail), the absence of a comb in the female, and the male's annual 'eclipse moult': from June to September, the male junglefowl develops a special temporary plumage, where the golden neck feathers are replaced with duller ones and the long tail feathers fall out. Domestic chickens have lost this feature somewhere along the way.

Junglefowl also have different flock dynamics, a short breeding season and a distinctive crow - again, features that would be invisible to an archaeologist. Unlike the heftier skeletons of larger farm animals such as horses or cows, fowl bones don't survive well in the archaeological record or often get carried off by scavenging animals. Even when bones are discovered, they can be difficult to differentiate from other similar species. When, in the 1980s, archaeologists discovered bones that looked remarkably hen-like on an 8,000-year-old Chinese excavation, it was initially hailed as solid evidence of the beginnings of chicken domestication. Later analysis, rather disappointingly, revealed the bones to be pheasant.[22]

Thankfully, when a species becomes domesticated, it's not just the behaviour of an animal that changes. The process of selecting and breeding animals, based on traits that are useful to humans, actively changes the genetics of that species. In other words, domestication leaves its trace in the genes. Only by looking at the minute differences in the genetic sequences of junglefowl and modern chickens can scientists identify the point at which one species evolved into the other. A large international team of researchers recently managed to collect and sequence over eight hundred genomes, including samples from chickens and all four species of wild junglefowl.[23] The study showed that the originator of *all* modern chicken breeds was a subspecies of the red junglefowl, the Burmese red junglefowl (*Gallus gallus spadiceus*), a bird whose home turf straddles south-western China, northern Thailand and Myanmar. The data also showed the domestication event took place between 12,500 and 6,000 years ago. Scientists had finally managed to establish the time period when wild junglefowl and the human race became inextricably linked, a journey that would take both species to unexpected, and often difficult, places.

The methods by which humans managed to tame the shy junglefowl are yet to be fully understood. The emergence of farming and the domestication of animals are often thought of as mutually dependent – it seems sensible to assume that people would only start keeping livestock when they had transitioned from hunting and gathering to living a settled life, growing crops in one location. The timing for chickens, however, is curious. Evidence for early farming in south-east Asia – in particular, millet, tapioca and rice cultivation – suggests that an agricultural way of life didn't emerge in the region until roughly 2000 BC, at least two thousand years too late to coincide with the domestication of the chicken.

A possible answer may be that the first relationship between junglefowl and human was not one of farmer and food source but of pet and owner. 'Pets' are often thought of as a uniquely modern phenomenon, an indulgence only possible with plenty of free time and spare resources, but humans have been keeping animals as companions and captives since time immemorial. When European voyagers began to encounter indigenous communities during their explorations of Asia, Africa, the Americas and Australasia, they were astonished to find hunter-gatherer communities - whose way of life hadn't changed for thousands of years - keeping animals for non-food purposes. From dingoes and wallabies in Australia to wolves, moose and bears in North America, a menagerie of creatures were being 'looked after' by humans, simply for the pleasure of it. Polynesians and Micronesians kept dogs, pigeons, parrots and fruit bats,[24] while in South America, particularly in the Amazon basin, the nineteenth-century English explorer Henry Walter Bates recorded at least twenty-two species living tame in local communities including deer, tapirs, monkeys, sloths, opossums, foxes, coatis, ocelots and jaguars.[25] Birds were also found to be valued pets, especially those that could sing or - like the junglefowl - had bright, decorative plumage. The domestication of a wild species, it appears, wasn't dependent on people living a settled, agricultural way of life.

The wild junglefowl is notoriously skittish but does have one trait that humans may have been able to exploit. Precocial birds - those that are hatched in an advanced state of development and can walk and feed themselves from birth - are also liable to wander off from their mothers. Imprinting - whereby a young chick bonds to the first thing it sees - keeps baby bird and parent in close proximity and helps it learn new behaviours.

The chicks of domestic fowl are well known for their ability to imprint on other animals, humans and even inanimate objects. Writing in the 1950s, the American zoologist Nicholas Collias noticed how Malaysian villagers 'often keep all sorts of pets including jungle fowl and perhaps this habit initiated the process of domestication. On the day of hatching, chicks of the domestic fowl have been shown to have a strong tendency to follow any large moving object, such as a person, especially if he talks or utters low-pitched, brief repeated sounds.' Collias went on to copy this process with a newly hatched, captive red junglefowl chick back in America.[26] Ernest Hubert Newton Lowther, a photographer working and living in India in 1949, also recorded two red junglefowl chicks in the wild following their human 'parent' straight after hatching.

If imprinting initially allowed humans to get close to wild junglefowl, it would have only been the beginning of the road to full domestication. Ethnographers working in the middle of the twentieth century documented different indigenous groups in the Malay Peninsula and western Thailand taking the eggs of junglefowl and hatching them.[27] As these junglefowl matured into adults, however, most of them wandered off back into the forest to reunite with their flock. Only a few birds, which seemed not to mind living in proximity to humans, stayed behind. Domesticating red junglefowl, therefore, may not have been simply a case of imprinting chicks or keeping wild birds captive. Wild animals that are held against their will – tethered or trapped, caged or corralled – tend not to thrive. Evolutionary biologists have long suspected that 'reduced fear' is the key trait needed for a wild animal to become domesticated and survive long term or, in other words, a genetic predisposition that allows them to tolerate being in contact with humans.

This theory of domestication was recently reproduced in a laboratory setting. In an intriguing project at Linköping University in Sweden,[28] red junglefowl were selected, over six generations, on the basis of how fearful they were of human contact. From around sixty male-female pairs of red junglefowl, the team bred hundreds of birds. Each successive generation of chicks was divided into groups based on how well they coped with a person handling them. The experiment soon demonstrated that the junglefowl who were selected for having a low fear of humans produced offspring that also had a low fear of humans - fearlessness was hereditary.

Perhaps more interesting, however, was the discovery that, over the generations, the offspring of fearless junglefowl began to display other traits. The less fearful birds not only ate more grain but also laid larger eggs than their anxious cousins - both characteristics that we see in modern chickens. Most surprising of all, however, was that the junglefowl who easily tolerated human company behaved more aggressively with their own kind. In other words, the birds who were confident around humans also tended to be bolder with their own kind, displaying more hostility towards their fellow flock members.

This pugnacious streak was a key discovery. Archaeologists had long suspected that the chicken's ancestor wasn't initially domesticated for eggs or meat. The wild junglefowl has a short breeding season, in spring, and the female will only lay four to six eggs in a clutch. As we'll see in Chapter 6, the daily laying habits of modern chickens wouldn't arrive for thousands of years. The desire for a reliable source of meat may have been a reason for humans to domesticate the red junglefowl but evidence of the breeding and eating of these early chickens is proving tricky to find. A recent archaeological dig in Israel, dated to the last two centuries BC, is one of the few ancient sites to provide evidence of their being eaten in significant numbers and is thousands of years after the initial domestication of chickens. Absence of evidence isn't, of course, evidence of absence, but it does raise an interesting question: why bother domesticating the junglefowl at all if you're not going to use them for food? The answer, archaeologists suspect, may lie in a surprising place. The cultural and historical journey of the chicken probably started not on the dinner plate, but in the fighting ring.

2

FIGHTERS

In the Wing

Malay Game Fowl

In the 1970s, an ancient burial ground was excavated on the outskirts of Vienna, Austria. There, archaeologists found more than seven hundred graves of Avar men, women and children. The Avars were a formidable people, warriors and herders who had originated on the Asiatic steppes and migrated west in the sixth century, enthusiastically looting and ransoming captives along the way. Glorious in life and even more glorious in death, the Avar burials revealed a treasure trove of exquisite weapons, clothes and jewellery. But the most interesting discovery of all was that nearly half the graves contained chickens, or parts of chickens.

At first, many dismissed the findings as 'food offerings', a common practice where the dead were given provisions for their journey into the afterlife. Later analysis of the bones and the burial site, however, revealed a remarkable pattern. The burials were 'gendered' – the men were interred with cockerels and the women with hens. Moreover, analysis of the skeletal remains of both the humans and their fowl companions revealed a remarkable dietary correlation between them. The chickens that had enjoyed a good diet in life were buried with humans who had eaten similarly well. High-ranking individuals had well-fed chickens. At the other end of the scale, lower-status humans were interred with birds that had enjoyed less nutritious meals.

Many of the high-status cockerels were also surprisingly old (the spur length and other skeletal features can be an indicator of age), suggesting that they had been given plenty of fuss and attention by their owners. One cockerel in particular appeared to have eaten better than many of the humans in the cemetery. This strange correlation told the archaeologists one thing: burials often reflect the value people put on certain objects and animals in their daily life – the Avars clearly had a special relationship with their chickens. Quite what that relationship was is difficult to infer but it didn't look like one of farmer and food.

We know precious little about how the chicken managed to colonise the world after its domestication in south-east Asia. It's a bird that doesn't go far, given the choice – chickens don't fly long distances and, although they're briefly buoyant in water, certainly aren't designed for swimming.* The chickens' ubiquity across the globe must be down to humans and humans alone. What's more, initial human interest in the bird seems to have been directed towards cockerels, not hens. And, while it's been difficult to determine what these male chickens were used for, many archaeologists suspect that the cockerel – with its natural fighting instincts and colourful plumage – might have

* Chickens aren't swimmers. They lack the webbed feet that would help them propel smoothly along, but it's their feathers that truly thwart any watery ambitions. Chickens are light, and will bob in the water for a few minutes, but in the absence of any waterproofing, their feathers soon become heavy and they sink like a stone.

attracted more interest as a ceremonial, ritual or fighting bird than a source of protein.

Excavations in modern-day Pakistan unearthed many treasures from one of the earliest major cities of the Indus Valley civilisation, Mohenjo Daro. Of those, some of the most interesting are 'stamp seals', small carved tablets that were used to make an impression into wet clay. No one knows for certain what their purpose was but archaeologists believe that seals were a way for traders to mark ownership of their goods or signature stamps for important people. Either way, seal stamps are often adorned with formidable creatures, divinities and mythical beasts as a way of denoting status and, perhaps, also serving as a powerful amulet. The animals on the stamp seals tell us something about Mohenjo Daro's people and their beliefs: it's significant that most of them are powerful beasts - bulls, rhinos, elephants, tigers, crocodiles and, crucially for us, cockerels. Linguistic scholars also speculate that the city's ancient name was *Kukkut arma* or 'City of the Cockerel', a testament to the male birds' prominence and perhaps spiritual significance.

The finer details of the chickens' epic journey from south-east to west Asia, into the Mediterranean and on towards northern Europe remain unclear. Archaeological evidence suggests that chickens had arrived in the Near East (Iran, Syria, Anatolia) by the third millennium BC and reached western Europe around 1000 BC, perhaps carried along coastal trading routes by the sea-faring Phoenicians.[1]

Evidence suggests that in these early days, the chicken - where it managed to make its presence known - was an exotic rarity. In ancient Egypt, whose inhabitants were assiduous in their documenting of all the creatures they farmed, ate and hunted, the chicken is conspicuous for its near absence until the fourth century BC, after which – as we'll find out later – Egyptians started to raise poultry for meat and eggs. Any references to chickens before this date are scarce. One of the few, and earliest, is found on a tiny scrap of limestone called an 'ostracon'* found near the tomb of Ramesses IX: a simple but lively sketch of a stylised cockerel dated to the twelfth century BC. Another, from around the same time, is an inscription mentioning the bird coming to Egypt as tribute, a special gift given as a sign of allegiance and submission from somewhere near modern-day Syria.[2] Such an offering would have only been prestigious if the chicken wasn't an everyday bird.

By the time of the ancient Greeks, however, cockfighting was an established, popular and hugely symbolic pastime, its image plastered on coins, pottery and monuments. Greece was a civilization that revelled in conflict and competition, a predilection it shared with the strutting, sparring bird. The Athenian general Themistocles is famously said to have had his spirits lifted by the sight of two cockerels scrapping. About to go into a decisive battle against the Persians, Themistocles

* In ancient civilisations such as Egypt and Greece, pieces of broken pottery or limestone were often used by everyday people as 'notepads' because papyrus was expensive and tricky to make. These shards were used for lots of different purposes – making notes, sketches, calculations, writing spells or votive offerings. The Greeks called them *ostrakon* and used them as ballot papers to decide whether a person should be banished from society as a punishment. This is where we get the modern word 'ostracised' from.

'espied two cocks fighting and immediately caused his army to behold them, and made the following speech: "Behold, these do not fight for their household gods, for the monuments of their ancestors, for glory, for liberty, or the safety of their children, but only because the one will not give way to the other." This so encouraged the Grecians that they fought strenuously and obtained a victory over the Persians.'[3] In the subsequent battle, Themistocles' soldiers gave the Persians such a pasting that cockfighting became emblematic of military valour and victory at all costs. Writing several centuries later, Pliny the Elder compared the Greeks' love of cock-fighting with the Roman passion for gladiatorial combat: 'At Pergamus, there is every year a public show of fights of game-cocks, just as in other places we have those of gladiators.'[4] Another writer, Aelian, described an annual cockfight held in remembrance of Themistocles' victory.

The Greeks associated the cockerel with Ares, god of war, and Athena, defender of Athens. For them, the brutality and thrill of mortal combat was not only entertaining but represented everything a male citizen should aspire to. Young men were obliged to watch cockfights 'for the sake of instruction',[5] while the philosopher Chrysippus marvelled at the usefulness of cockerels for 'inciting soldiers to war and instilling an appetite for valour'.[6] Even the Greek word for cockerel, '*alektor*',* meant 'defender' or 'averter of evil'.[7]

In his work *Anacharsis*, the satirist Lucian, writing in the second century AD, gently poked fun at the Greek obsession

* The Greek word lives on in 'alektorophobia', the fear of chickens.

with the violent confrontation played out in the cockfighting pit: 'What would you say, if you saw our quail- and cockfights and the not inconsiderable zeal we devote to them? Or is it likely you would laugh, and especially if you learned that we do it by law and that all men of military age are instructed to attend and watch the birds flail at one another until their very last fall? But it is not ridiculous, for an appetite for danger steals gradually into their spirits so that they might not appear less noble and daring than cocks and give in while they still have life under distress of wounds and exhaustion or some other hardship.'[8]

Art, sculpture and armour also used the male chicken as an explicit motif for courage - shields carried by the Greek infantry were emblazoned with images of cockerels, while special amphorae, given out as prizes at the Panathenaic Games, were adorned with cockerels as a symbol of an unquenchable fighting spirit. Equally, the defeated cock - if it survived the fight - was believed never to crow again, a vanquished failure. Submission was thought the worst possible outcome of all; the Greek phrase 'like a beaten cock' meant a fate worse than death - a life of slavery.

For the Greeks, violence and male virility were comfortable bedfellows. Contests between two cockerels were loaded with symbolism - not only about male competition but also the proximity of love and pain. The cockerel also represented four of the traits Greek men most highly valued in themselves - 'pugnacity, pride, sexual appetite and alertness'.[9] Most importantly, the cockerel's perceived promiscuity and dominance of its partner mirrored one of ancient Greece's more peculiar practices - sexual relationships between adult males and boys. Far from being viewed by society as subversive or abusive, 'pederasty' was seen as a key rite of passage in an adolescent male's social and military training, especially those from the upper classes. The balance of sexual power between the adult and the youth was critical, however. The distinction between the active 'pursuer' role of the older man and the passivity of his young lover was important and the cockerel proved an apt metaphor for pederasty in Greek sculpture and ornaments. Adult suitors would often give their young male amours a piece of jewellery or ornament depicting a cockerel, a sure-fire sign of a dominant and energetic lover.

In fact, the Greeks were so convinced of the cockerel's virility that both scholars and shamans heartily recommended chicken testicles as a powerful aphrodisiac. The elder Pliny's *Naturalis Historia*, which relied heavily on earlier Greek texts for much of its content, confidently prescribed 'wearing the right testicle of a cock, attached to the body in a ram's skin' as a powerful male stimulant, although - he added - the charm had the opposite effect if you rubbed the testes with goose-grease first or popped a vial of cockerel blood underneath the bed.[10]

There were very few things, it seemed, that a good cockerel spell couldn't cure. Pliny went on to recommend 'the flesh of cocks' or 'the brains, taken in wine' to neutralise the venom of serpents, while 'panthers and lions will never touch persons' who have been rubbed with cockerel soup. The addition of a pinch of garlic, insisted Pliny, made the charm practically fool proof. Moreover, such was the potency of cockerel flesh that if it was 'mingled with gold in a state of fusion, it will absorb the metal and consume it'.[11] In another particularly unpleasant spell, the sufferer is advised to rip the tongue out of a live cockerel to silence his enemies.[12] Cockerel charms might also be used to protect a vulnerable harvest. Pausanias, Greek traveller and geographer of the second century AD, noted a curious local ritual to protect grapes from frosty, bud-nipping gales: 'So while the wind is still rushing on, two men cut in two a cock whose feathers are all white, and run round the vines in opposite directions, each carrying half of the cock. When they meet at their starting place, they bury the pieces there.'[13]

The Romans were no less dazzled by the fighting prowess of the cockerel. Emperor Septimius Severus, who ruled from AD 193 to 211, was notorious for his brutality. When he travelled across Britain in 208, with the intention of conquering Caledonia in the north, he famously issued his troops with the order: 'Let no one escape sheer destruction, no one our hands, not even the babe in the womb of the mother, if it be male; let it nevertheless not escape sheer destruction.'[14] Keen to harden up his two young sons, Caracalla and Geta, for the demands of imperial expansion and warfare, he encouraged them to keep their own cockerels and pit them against each other in the fighting ring. According to the Roman writer Herodian, who penned a gripping account of the history of the empire, the bird battles failed to have the desired effect and instead only served to fuel the bitter rivalry between the two brothers.

It's perhaps no surprise to learn that, after a childhood dominated by a sport that celebrated inter-male violence, as an adult Caracalla not only attempted a coup but also tried to persuade doctors to kill his father. On Severus's death, Caracalla seized control and set to work murdering anyone who remained loyal to his late father or his brother. And although Severus had intended his sons to rule together, Caracalla couldn't rest until his victory – like a fighting cockerel – was absolute. Only a few months later, he had his brother murdered and claimed sole charge of the Roman Empire for himself.

Indeed, the Romans were so assured of the cockerel's courage and strength that they clamoured after *alectoria gemma*, or cock's gems. These were small, polished stones found in the gizzards of cockerels; chickens often swallow stones or pieces of grit to help break down their food and aid digestion. Over time, these stones are polished smooth by the action of the animal's gastrointestinal tract and then regurgitated or passed. For the superstitious Roman, the older the cockerel, the more powerful the cock gem would be once it was retrieved. By holding the alectory in one's mouth or hand, or swallowing it whole, the lucky recipient would experience a rush of vim and vigour, especially if a man was hoping to boost his sexual athleticism or a woman her feminine allure. Pliny the Elder wrote: 'Especially in matters of sexual desires it renders [the male] vigorous, strong and robust. But it will even be of help to women carrying it who wish to please men.' [15]

Milo of Croton, one of antiquity's most renowned beefcakes and six-times wrestling champion at the ancient Olympic Games, was said to have carried a cock gem about him, 'a thing that rendered him invincible in his athletic contests'.[16] The cock gem appeared to work; Milo's death, when it finally came, wasn't at the hand of an opponent. In what has to be one of the strangest demises in history, according to legend, Milo was attempting one of his famous demonstrations of strength by pulling a tree trunk apart with his bare hands. In a cruel twist of fate, the wedges that had been used to make an initial split in the tree fell out and the trunk closed upon the wrestler's hands. Unable to free himself, Milo waited for help, but none was forthcoming. As Milo stood waiting, no doubt wondering whether the potency of his cock gem had finally run its course, he was set upon and devoured by a pack of wolves.

The male chicken was also central to another of ancient Rome's bizarre magical rituals. Alectryomancy was a method of foreseeing the future by watching poultry peck at grain. In a chickeny version of Ouija, a cockerel (preferably white, for its perceived purity) would be charged with the task of answering life's tricky questions – matters of public policy, for example, or the success of a forthcoming marriage. Before a major decision was made, the different letters of the alphabet would be drawn in the sand and then a piece of grain placed upon each. The diviner would then make a note of the order in which the cockerel ate the grain, carefully jotting down the letters. The way a chicken gobbled its food was also auspicious - records show that cockerels were often taken into the theatre of war as a way of predicting the outcome of a battle. Pliny tells us that 'there is not a mighty Lord or State of Rome, that dare open or shut the Door of his House, before he knows the good Pleasure of these Fowls [...] These birds command those Great Commanders of all Nations upon Earth.'[17]

Publius Claudius Pulcher, Roman politician and commander of the only fleet to suffer a major defeat in the First Punic War, was tried for incompetence and severely fined for ignoring the advice of his chickens. When the birds on board his ship refused to eat - a sure sign that defeat was imminent - Publius is said to have picked up his fortune-telling poultry and hurled them into the waves, shouting *Bibant, quoniam esse nollent!* - 'Since they do not wish to eat, let them drink!' Publius was lucky, however, to have dodged a more unpleasant ruling. Sentencing for crimes in the Roman legal system often involved cruel, emblematic punishments. A person guilty of parricide, murdering a parent or other near relation, would be sentenced to *poena cullei* or 'penalty of the sack'. The perpetrator would be first whipped or flogged and then sewn up into a sack with three live animals, before finally being thrown into deep water to drown. The creatures of choice were a dog, a cockerel and a viper. While the snake was seen as a creature of the underworld, a place to which the criminal would surely be heading, the poor dog and cockerel - both animals prized for their vigilance against attack - were also punished for failing to alert the criminal's victim.[18]

Chickens or, more exactly, the birds' feathers, were put to ingenious use by the defenders of the Greek town of Ambracia during a siege by the Roman general Marcus Fulvius Nobilior. Nobilior's troops were battering down the walls surrounding the town, in 189 BC, but as quickly as the soldiers made progress, the town's resourceful inhabitants were hastily plugging the gaps with debris. Enraged, the general ordered his men to start digging a tunnel underneath the perimeter walls. The townsfolk, however, had got wind of the plan and begun to dig their own tunnel, aiming to meet the Romans face to face underground. When the two tunnels finally touched, spears flew but neither side could break the other's defences, and so the Greeks came up with an even more cunning ruse.

They filled a huge terracotta jar with chicken feathers, dragged it into the tunnel and set the plumage alight. Great clouds of thick, acrid smoke swirled from the pot, which the Greeks then wafted down the tunnel using large blacksmiths' bellows. As contemporary writer and historian Polybius described in his *Histories*, 'The plan was successfully executed; the volume of smoke created was very great, and, from the peculiar nature of feathers, exceedingly pungent, and was all carried into the faces of the enemy. The Romans, therefore, found themselves in a very distressing and embarrassing position, as they could neither stop nor endure the smoke.'[19] Chicken feathers are rich in toxic sulphur, so the smoke produced would have left the Roman attackers breathless and choking. Although the residents of Ambracia surrendered to the Romans in the end, their tactics were not forgotten. It was one of the earliest records of chemical warfare in human history.

When Julius Caesar wrote *De Bello Gallico* during his ferocious campaign to suppress the tribes of northern Europe, he made some interesting observations about the Celtic tribes in Britain: 'They do not regard it lawful to eat the hare, and the cock, and the goose; they, however, breed them for amusement and pleasure.'[20] From this brief comment, much has been inferred. While it's clear that the chicken had already reached British shores before Caesar's troops landed in 55 BC, quite what 'amusement and pleasure' meant is less clear-cut. Some historians have interpreted this as a reference to cockfighting, others that the bird held some kind of ceremonial or ritual purpose.

The archaeology helps us fill in the gaps. Zooarchaeologists - people who study ancient animal remains - have studied poultry from this period. Dating these very old chicken bones has helped pinpoint the birds' introduction onto British soil at sometime between the fifth and third century BC, at least two hundred years before Caesar's invasion. How the chicken reached our island, well before the Romans, isn't clear; one theory is that the Phoenicians might have brought chickens to south-western Britain to trade for tin, one of the necessary ingredients in the production of bronze. Chickens are also fairly rare in Iron Age excavations and those that have been found are often buried with remarkable care, with their skeletons intact. The usual signs that a chicken has been eaten - such as evidence of butchery or gnaw marks - are also absent. Analysis of bones reveals that the age of death of Iron Age chickens – rather like the Avar birds – is surprisingly old. A study by the University of Exeter found that chickens were living for two, three, even four years – a stark contrast to the brief two months of a modern chicken destined for Sunday lunch or a box of nuggets.[21]

From all this information, people are starting to piece together the chicken's role in Iron Age Britain. It seems unlikely that the chicken, when it first arrived, was a dinner-time staple. Iron Age sites across Britain, and other parts of northern Europe, often reveal an interesting ratio of cockerels to hens, one that wouldn't be found if the birds were being raised primarily for meat or eggs. At three to one, cockerels outweigh the presence of females to such an extent that meat or egg production has been ruled out by most experts.

Cockerels don't tend to last long when chickens are kept primarily for food. A flock of hens only needs one cockerel to fertilise its eggs, and too many males can lead to violent battles. In societies where eggs are used as a food source, the male bird is largely redundant and either dispatched as a chick or eaten within a year. Equally, where chickens are reared for meat, old cockerels and hens make tough eating and so aren't traditionally kept for long periods of time. When archaeologists find cockerel bones that don't have any signs of butchering or gnawing, or find skeletons of old male birds, all the signs point towards a culture that valued the birds not for their tastiness but for their symbolism, rarity or sport. A rare Iron Age coin found near Chichester in West Sussex, displays an intriguing image of a chicken with a man's face; such a strange hybrid creature has been interpreted as a sign that Celtic tribes, just as the Greeks and Romans did, were exploring ideas about male identity and combat.[22] The coin, although the only one discovered in Britain, is remarkably similar in design to others found across the water in northern France, indicating cross-channel influence or a shared culture.

It's also possible that Iron Age people associated the cockerel with a particular deity. The Greeks, Romans and native tribes of northern Europe all shared a common belief in multiple gods, each with its own set of skills or attributes. While the names of these deities differed between cultures, their characteristics were often similar. The Romans, for example, believed in Mercury, a god born at dawn, a divine messenger and guider of souls to the afterlife. In Roman mythology, Mercury's totem animals were a ram, symbolising fertility, and a cockerel, a similarly lusty animal that also announced the break of day. One of the more moving discoveries of recent years was an exquisite bronze figurine of a cockerel found in a young child's grave from Roman Cirencester, in Gloucestershire. The toddler had been lovingly buried wearing tiny hobnail shoes and was accompanied by a pottery 'sippy cup' and the small cockerel. As the emblem of Mercury, archaeologists believe the cockerel was placed in the coffin by a bereaved parent in the hope that it would keep the child safe on its journey to the afterlife.

Caesar, in his observations of the tribes of Britain, noticed that the native people seemed to believe in a similar god to Mercury. What the Iron Age people called this deity isn't clear but Caesar saw close enough similarities that he felt confident enough to call it by the Roman name; 'Amongst the gods,' he wrote, 'they worship Mercury above all and he is the one with the most numerous representations.'[23] And while Caesar was imposing his own interpretation on a Celtic god, it seems there was enough of a resemblance between the two deities for one to eventually subsume the other over time. Whatever symbolism the cockerel did or did not hold for the people of Iron Age Britain, after the Roman conquest of the first century the bird became a recognisable and potent emblem across many parts of the country. Native Britons lived under Roman rule for three and a half centuries and, during this time, many Roman and Iron Age artistic and religious practices became intertwined. This period of synthesis between the conquering nation and the native tribes is often called the Romano-British and the cockerel enjoyed a long period as a popular motif, synonymous with Mercury and his attributes. The male chicken, brimming with such religious significance, also became the ideal bird for sacrifice.

In 1976, workmen digging a pipe trench for Severn and Trent Water Authority accidentally stumbled upon the remains of a temple on the edge of the village of Uley in Gloucestershire. Over the next two years, archaeologists uncovered a site that had been used as a place of ritual worship for thousands of years. For successive generations, the space had held a special fascination for local people – Neolithic tribes had erected standing stones, followed in turn by an Iron Age enclosure with two timber shrines. When the Romans finally arrived in this sleepy corner of the Cotswolds, they built a stone temple on the site and dedicated it to Mercury. What was astonishing about the site was the sheer number of animal remains it contained - around a quarter of a million bones - including an unusually high proportion of rams and cockerels. The sex and age of the animals was hugely significant - the sheep were male, young and in prime health. The chickens were also specially selected - and male - and most of the sacrifices seemed to have happened in autumn, perhaps in honour of a seasonal festival now lost in time.

The ritual killing of thousands of healthy rams and cockerels, animals that embodied the attributes of Mercury, was a way of creating a dialogue between local Romano-British people and this important god. It must have also been expensive - and time-consuming - for the community who raised the livestock. But as a means of honouring and propitiating Mercury, and thereby ensuring that the god would continue to smile on them, it was clearly seen as essential. It's almost impossible to imagine how powerful and enduring this belief system was. Not long after the temple was built, a human-sized statue of Mercury was also carved from local Cotswold limestone and placed on a base with a stone ram and cockerel as his companions. When the Romans began to lose their grip on the region, at the end of the fourth century, the statue of Mercury was toppled and broken into pieces. Interestingly, whoever smashed the effigy must have believed in the power of Mercury or the spiritual significance of the location; they carefully decapitated the statue and buried his head, along with the ram, under a cobbled platform surrounding the building. They also carefully placed the legs and body of the stone cockerel in the foundations of the building that succeeded the Roman temple. Such care and deliberation with Mercury's remains suggest that whoever did it still believed in the potency – either for good or evil – of the statue and his stone companions.

The cockerel was also associated with another Roman god, Mithras. Mithras is thought to have started life as an Indo-Iranian deity, a god associated with loyalty and justice, but at the height of the Roman Empire his worship was transformed into a secretive cult. Although we know very little about its practices, archaeologists believe Roman Mithraism was exclusive to men, a religion that celebrated allegiances and mutual obligation. These notions particularly appealed to military personnel; emperors and officers at first, but soon also the rank-and-file soldiers, especially those stationed at the borders of the Roman Empire.

Although the exact nature of Mithraic rituals is poorly understood, there are a number of sites where cockerels seemed to have been sacrificed in surprisingly large numbers. At a temple in Tienen, in Belgium, archaeologists found evidence of a single sacrifice of almost three hundred cockerels, while in temples dedicated to Mithras in London, the chicken was the species that most frequently got the chop.[24] Quite why this was, isn't clear. The pugnacious nature of the cockerel might seem to provide a plausible reason for it playing a lead role in the rituals of a military cult, but there is another possible explanation. The original Indo-Iranian Mithras was also the god of light and the sun, so the sacrifices may have been linked to the significance of the cockerel's crow at daybreak.

The spiritual association of the cockerel endured throughout Britain's Roman occupation, but the empire was also beginning to value the bird as a food source. Archaeological and documentary evidence shows that the Romans became rather good at chicken farming and their techniques soon spread across conquered territories. Vindolanda, a Roman fort in Northumbria, just south of Hadrian's Wall, has yielded more than a thousand thin, postcard-sized tablets, whose contents provide historians with fascinating insights into everyday life in a far-flung corner of Roman Britannia. From party invitations to military orders, personal messages to requests for beer, these rapid scribblings from the first and second centuries also include shopping lists for food from local markets and suppliers. One tablet reads 'chickens, twenty [...] a hundred or two hundred eggs, if they are for sale there at a fair price [...] 8 sextarii* of fish-sauce [...] a modius of olives'. Chickens and eggs, it appears, were on the menu.

Columella, a prolific agricultural writer in the first century AD, dedicated a huge section of his work *De Re Rustica* ('On Farming') to chicken farming. In it, he gave a wide and detailed description of the new art of breeding and keeping chickens for 'economic return'. Columella noted that, while the Greeks raised birds primarily for their large body size and aggression,

* A sextarius was a unit for liquid, about the equivalent of half a litre or a pint.

the Romans were starting to expand their efforts to breed chickens for egg-laying and meat production. With not a little disapproval, he noted: 'we lack the zeal displayed by the Greeks who prepared the fiercest birds they could find for contests and fighting. Our aim is to establish a source of income for an industrious master of a house, not for a trainer of quarrelsome birds, whose whole patrimony, pledged in a gamble, generally is snatched from him by a victorious fighting-cock.'[25] Certain breeds, notably those originating from Rhodes and Persia, were heavy, strong fighters but poor layers. Roman farmers were beginning to experiment with crossing cockerels and hens from different regions in the empire and other trading nations. Traits such as broodiness and the ability to fatten up quickly were high on the wish list, as were more aesthetic 'show' points such as appearance and bearing.[26]

It's surprising just how commercially-minded Roman chicken farmers could be. Columella described the importance of well-ventilated housing, the correct balance and nutrition of feed, culling, disease management and even business practice. With typical lethal efficiency and lack of sentiment, Roman farmers also perfected the art of 'cramming', a method of fattening birds by urban chicken keepers or 'poulterers'. Young birds were raised in largely free-range conditions in the countryside but were sent into the city to be plumped up in finishing units. There, rather like the geese or ducklings raised for *foie gras*, the young chickens would be stuffed with frequent meals of softened meal or bread and kept in the dark in small, tightly packed enclosures for three or four weeks until slaughter.[27]

Many Roman techniques seem remarkably 'modern', such as maximising the productivity of layers through artificial selection and culling, sending poor layers and older birds for slaughter; and keeping detailed records on the productivity of individual birds. Some farmers even learned how to identify and artificially incubate fertilised eggs and developed covered drinking troughs, which kept the chickens' water clean and fresh.

While many of the practices made sense in terms of animal husbandry, the Romans also couldn't resist bringing their superstitious beliefs and bizarre folk remedies into the poultry yard. Columella was particularly concerned about 'pip', a catch-all term for various respiratory diseases in chickens that cause them to breathe heavily through their mouths. He recommended an unusual cure: 'wet their mouths with warm human urine and keep them closed until the bitter taste of the urine forces them to expel through their nostrils the nauseous matter produced by the pip'.[28] Thundery weather was thought to affect the taste of chicken eggs and kill hatchlings still in their shells; 'lay a little grass', he advised, 'under the litter in the nest-boxes and small branches of bay and also fasten underneath heads of garlic with iron nails, all of which things are regarded as preservatives against thunder'.[29] Young chicks were believed to be particularly vulnerable to malignant forces; 'care must be taken that they are not breathed upon by snakes, whose odour is so pestilential that it kills them all off. This is prevented by frequently burning hart's-horn* or

* Heating up 'hart's-horn', ground deer antler, produced ammonia and carbon dioxide. It was also used by bakers in powder form to leaven bread before the discovery of bicarbonate of soda.

galbanum* or women's hair; by the fumes from all these things the aforesaid pest is generally kept away.'

Eggs were useful to the Romans not just as a food source but as an ingredient in paint. Egg tempera is a fast-drying medium for painting that mixes egg yolk with coloured pigments. The addition of egg, with its glutinous texture, not only gives paint a lovely workable consistency and adherence but also, when it dries, adds a hard-wearing lustre to the surface. The Romans perfected the technique and used it, along with other binders such as beeswax and wheat paste, to adorn their houses. Excavations of Roman dwellings across entire cross-sections of society reveal sumptuous decoration and brightly coloured wall murals in almost every room. And while we often think of *fresco* as the preferred method of Roman painting (where pigment is applied to wet plaster), many artists used egg tempera on dry surfaces, especially timber.

One of the most famous surviving examples of egg tempera, dating from around AD 200, depicts Septimius Severus and his young family, before his son's murderous spree. The 'Severan Tondo' was painted in happier times – Caracalla, still a small boy, and his younger brother Geta, no doubt painted in a spare moment between cockfights. After Geta's later assassination by Caracalla's soldiers, the painting was also mutilated. Geta's face was scratched away and smeared with excrement in what experts believe was an act of *damnatio memoriae*, an officially sanctioned 'scrubbing out' of any trace of Geta by supporters of Caracalla.

* Galbanum is the aromatic resin of a plant related to fennel.

Of course, the chicken didn't just travel west from its Southeast Asian homeland. Perhaps more remarkable than its journey through Near Eastern and Western civilisations was the bird's ambitious sea voyage eastwards across the unimaginably vast Pacific Ocean. Tracing the bird's presence as it hopscotched from remote island to remote island also helps scientists and historians map out the human story of seafaring expansion, one of the most fascinating but little understood parts of migration history.

Current evidence suggests that, around 3000 BC, fisherman-farmers, who were skilled at long-distance canoe travel, paddled their way boldly into the big blue void. These ancestors of modern Polynesians probably set out from Taiwan, colonising islands as they went and bringing with them the domesticated animals and plants they relied on for survival. By 800 BC their descendants had reached the islands of Samoa and Tonga, over five thousand miles away from their original starting point. After a thousand-year hiatus, a second wave of migration reached even further east, colonising Tahiti by AD 700, Hawaii two hundred years after that and New Zealand by AD 1200.

One of the biggest bones of contention is just how far east, across the Pacific, the ancient Polynesians and their chickens got. Easter Island, or Rapa Nui, is often considered the furthest point east reached by these intrepid pioneers. It's spectacularly remote, lonely in its isolation at over two thousand miles from the west coast of Chile. Sometime between AD 800 and 1200, canoe-bound men, women and perhaps even children landed on shore, bringing with them tools, knowledge, and the plants and animals they hoped to raise. Along with bananas, sugar cane and taro, they brought chickens and, either by accident or as a potential food source, the Polynesian rat.

Some archaeologists and geneticists, however, think they didn't stop there. It's long been debated whether the Polynesians carried on and made it all the way across the Pacific to the coast of South America, beating Columbus to the New World by hundreds of years. Two foodstuffs have been at the heart of the controversy: one is the sweet potato and the other is the chicken, a combination that has been dubbed the 'chicken and chips' theory of migration. The sweet potato – a food native to South America – had already found its way to the islands in the South Pacific well before Columbus; archaeologists have found prehistoric residues of sweet potato dating back to at least AD 1100.[30] Some experts have suggested that the sweet potato accidentally bobbed along the waves from the South American coastline to remote South Pacific islands, but others suspect the tuber may have been deliberately traded between cultures. The lexical similarity between the word for sweet potato in the ancient Polynesian tongue – *kuumala* – and the name for the plant among the Quechua people of South America – *kumara*, *cumar* or *cumal* – also suggests some kind of pre-Columbian contact between the two geographically separate groups of people.

The presence of chicken bones at an ancient archaeological site on the Chilean coastline also indicates that Polynesian people and native South Americans had met each other before Columbus arrived. For years, scholars had assumed that Portuguese or Spanish explorers brought chickens to the continent sometime in the early sixteenth century. When these ancient South American chicken bones were analysed, however, the results indicated that they dated from the late fourteenth century – at least a hundred years before the arrival of the Europeans – and that the chickens' DNA was similar to prehistoric chicken remains found on Tonga, Samoa and Easter Island.[31] What's fascinating, however, is that genetic studies of modern South American people have yet to uncover any indication of Polynesian ancestry. Landing on shore, our Polynesian explorers seem to have traded, dumped or gifted chickens to local Chileans and then turned around and gone home.

For researchers who understand the navigational skills of the Polynesian people, the suggestion that they might have made it all the way across to South America is utterly uncontroversial. If finding the needle in a haystack that is Easter Island presented few problems for these exceptionally skilled seafarers, discovering an entire continent's coastline would have been relatively uncomplicated. And if the Polynesians had set out from Easter Island, the nearest launch point, the voyage to the Chilean coast would have taken only a few weeks, a manageable journey for an experienced crew and their feathered cargo.

Polynesian sailors are thought to have passed on navigational knowledge through oral history and songs, and used the stars, rhythms of the ocean and natural cues to orientate their way over huge distances in outrigger canoes and double-hulled canoes with sails. In 1976, a crew successfully recreated and sailed a traditional Polynesian vessel from Hawaii to Tahiti, without a compass, sextant or any other navigational instruments. The boat was also loaded with the Polynesian 'farming package' in an attempt to see how the animals and crops coped with the voyage; along with a dog, pig, a cockerel and a hen, the boat also carried sprouting coconuts, breadfruit, sweet potatoes, sugar-cane roots and other species of plants thought to have been cultivated or used by early Polynesians. Under the able leadership of Captain Kawika Kapahulehua, the crew of seventeen men took just thirty-four days to successfully complete the 3,000-mile voyage and even less time – three weeks – to return. And despite a few frayed tempers and fights between crew members along the way, the dog, pig, cockerel and hen all survived both legs of the journey in rude health.[32]

Whether the chicken first crossed the Pacific for the purposes of cockfighting, eggs or meat (or a combination of all three) isn't clear. Cockfighting remains a popular and fiercely guarded pastime across many of the island communities, but the history of the sport is difficult to establish. The only clues to its antiquity come from the few written records of early European explorers and even those are relatively recent compared to the chicken's presence in the Pacific region. When the Italian scholar and explorer Antonio Pigafetta first encountered people living in the Philippines, in 1521, he was surprised to find a culture who raised 'fighting cocks and bet on their favourite birds'. 'They have large and very tame cocks,' he continued with admiration, 'which they do not eat because of a certain veneration that they have for them. Sometimes they make them fight with one another, and each one puts up a certain amount on his cock, and the prize goes to him whose cock is the victor.'[33]

William Marsden's *History of Sumatra*, first published in 1783 following his return from the region after thirteen years with the East India Company, noted that 'the passion for cock-fighting is so great, that it is rather a serious occupation among the inhabitants than an amusement. A man in that country is rarely met travelling without a cock under his arm; and sometimes there will be fifty persons in company with their cocks. They often risk everything upon the event of a battle, even their wives and daughters, and the loser is frequently stripped of his goods and reduced to despair.'[34] And yet chickens and their eggs were also being eaten on many of the Pacific Islands. In 1722, the first Europeans set foot on Easter Island. After the Dutch sea captain Jacob Roggeveen and his crew had finally made it on shore, blithely shooting a dozen or so islanders in the process, the native people anxiously attempted to restore peace with a feast. 'After the inhabitants of this Island had learned the power of our guns,' Roggeveen later boasted, 'they began to treat us very polite and kindly offered from their huts all sorts of earth fruits, sugar cane, Jambe Jambes,* bananas and a large number of chickens, which we liked very much and were a good refreshment.'

* Possibly yams.

Given that the Polynesians carried chickens across wide stretches of the Pacific, ferrying them to some of Earth's most remote islands, one mystery remained unsolved until recently. Ancient chicken bones are conspicuously absent in archaeological sites in New Zealand. If the Polynesians colonised New Zealand in the late thirteenth or early fourteenth century, and brought chickens along with them as they did elsewhere, why is there not more evidence of hundreds of years' worth of them keeping or eating poultry? Researchers recently attempted to solve the riddle by carbon-dating the few chicken bones that have been found at historic New Zealand sites.[35] The results were a surprise. The chicken bones were actually remarkably young, and from a very narrow time frame in the second half of the eighteenth century. This predates the permanent European settlement of New Zealand in 1840, when Britain annexed the islands, but is much later than the arrival of the Polynesians, whose ancestors went on to become the Māori people. The best and most exciting explanation is that the chickens were introduced by Captain James Cook and his ship, the *Resolution*, on their second expedition in 1773.

Cook often gave domestic animals as gifts to local tribal leaders. When Cook first landed on New Zealand at *Tūranganui-a-Kiwa* (which he later named 'Poverty Bay') on 8 October 1769, his initial encounter with the indigenous Māori people was a disaster. Within just a few days, the British had shot and killed a number of Māori, including *Te Maro*, a local chief. Keen not to repeat the same mistakes, on Cook's second and third expeditions to New Zealand he attempted to grease the wheels of contact with gifts of ornaments and animals. Boars and sows, goats and chickens were all brought ashore during Cook's successive landings. His journal entry for 3 November 1773 describes how he presented a Māori chief in Cloudy Bay with two cockerels and two hens, which 'he received with such indifferency, as gave me little hopes that proper care would be taken of them'.[36] In total, Cook gifted nineteen chickens across New Zealand that year but it's thought that more birds were either secretly sold or given away by the crew. Cook admitted: 'More Cocks and Hens are left behind than I know of as several of our people had of these as well as my self, some of which they put on shore and others they sold to the Natives, whom we found took care enough of them.'[37]

As well as using chickens as bargaining chips, Cook had another, more pressing motive for the release of domestic animals on New Zealand soil. While Cook's first voyage had been, for all intents and purposes, a scientific one sponsored by the Royal Society to observe the transit of Venus, subsequent trips were about securing New Zealand and other areas of the Pacific for British trade and exploitation. Cook's crews deliberately released pigs, goats and chickens into the wild with the intention that they would naturalise and provide a source of fresh food for future commercial visits.[38]

Cook hoped his birds would be well received, but from all accounts the Māori were decidedly unimpressed. And this perhaps gives researchers a clue as to why the Māori people hadn't bothered with poultry until after they encountered Europeans. Prior to the arrival of Polynesians, New Zealand was home to a spectacular and unique range of terrestrial birds, including various species of moa, a creature that could range in size from a large turkey to a 3-metre-tall, 300-kilogram titan. The moa was also defenceless, flightless and, unfortunately, delicious – a holy trinity of traits that made them irresistible to early Polynesian settlers. Even if the first Polynesian pioneers had brought a handful of chickens to begin the new colony, as they did with other islands, it would have soon become apparent that the land provided more protein than domestic poultry ever could; there was no need to breed chickens in the presence of such abundant wild alternatives.

Within two hundred years of Polynesian colonisation, however, much of New Zealand's larger game had been hunted to extinction. And while the islands continued to provide wild food in the form of sea birds, sea lions, seals, shellfish, fish and eels into the eighteenth century, Cook may have arrived at a critical point when the Māori were facing serious food shortages. Their initial ambivalence towards the chicken, which was understandable given the island's historic plenty, may have been quickly replaced by a realisation that poultry offered a potential lifeline. Exactly what happened to Cook's own chickens is not known. But his fears that they would perish within a short time may have been unfounded, as it seems that the Māori quickly turned to breeding and trading chickens between themselves. Other foods that were introduced by the Europeans – including potatoes, carrots and cabbages – also became central to the Māori economy and way of life.

One question, however, remains. When the Māori realised they had hunted New Zealand's large game to extinction back in the fifteenth century, being such experienced sea travellers why did they not re-engage in long-distance trade with other Polynesian communities and bring a new batch of chickens back to the island? All the archaeological evidence suggests that, after the middle of the fifteenth century, long-distance inter-island journeys across the Pacific dramatically declined. What happened to all those thousands of years' worth of navigational confidence and expertise? One idea is that the 'climate window', the perfect wind and sea conditions that had previously facilitated long-distance sea travel, suddenly closed, never to return. Without a fair breeze behind them, it seems the Polynesians and their poultry could no longer ride the waves.

The idea of a hen coping with life at sea, never mind the discomfort of a storm-tossed sea, might seem far-fetched, but there is a rather sweet ending to this nautical tale. In 2019, the *Observer* newspaper ran the headline 'Have hen will travel: the man who sailed around the world with a chicken'. Between 2014 and the end of 2018, young French explorer Guirec Soudée and his pet hen, Monique (or Momo for short), had been sailing the high seas on a small one-man boat. Starting in Brittany, Soudée and his feathered shipmate covered 45,000 miles, crossing the Atlantic and visiting both the North and South Poles, before returning home. Soudée had decided he wanted companionship on his lonely adventure and chose Monique – a Rhode Island red – as he thought she would provide not only comfort but a steady stream of eggs.

With a coop on deck and another in the cabin, Monique could sleep, lay and stay dry in the worst of the weather, but ended up spending most of her time with Soudée, wandering up and down the boat and calmly watching the world go by. At various stop-off points, Soudée and Monique would stretch their legs on dry land and take in the sights. In Greenland, the pair even overwintered for four months, trapped in the polar ice; undeterred, Soudée made Monique a tiny sledge and a sweater from his woollen gloves, and they went exploring, seeing Arctic foxes, caribou and the Northern Lights along the way. Monique also provided her companion with over a hundred eggs, a feat that probably saved Soudée's life when provisions ran perilously low. Once the ice melted, Soudée and Monique drifted onwards, becoming the youngest navigator and first chicken ever to sail the North-West Passage. The Polynesians would have been proud.

3

IDOLS AND ORACLES

The Hen Commandments

Dorking Chicken

In 1961, workmen were making repairs on the first floor of Lauderdale House, a magnificent Tudor home in London's leafy Highgate. As they began work near the chimney breast, carefully removing bricks from around the fireplace in the Long Gallery, the builders discovered a small secret compartment that had been bricked up over four hundred years earlier. In it, they found a drinking glass, two odd shoes, a candlestick and a length of cord. Alongside this seemingly disparate array of household objects were the bodies of four chickens and a single egg, untouched since the day they were placed there. Later analysis showed that two of the chickens had been strangled prior to their internment and two were bricked up alive. One of the chickens may even have laid the egg while it awaited its ghastly fate.

This grim collection is thought to be a 'counter-charm', designed to ward off evil spirits or witchcraft. Dozens of examples of these domestic sacrifices have been found in ancient buildings; the objects are often worn-out shoes, metal tools or scissors, but cats, birds and other small animals were also secreted in walls or under floorboards. It's difficult to get inside the mind of the medieval householder but objects selected for house protection spells are thought to have been chosen for a number of different reasons. Some materials, such as iron, were thought to be particularly effective against malevolent forces. Other attributes, such as an object's sharp edges, were believed to act as a physical deterrent. Some items, such as single shoes, were traditionally associated with fertility and infant health, while certain creatures were believed to be witches' companions or 'familiars'. Through death, it was hoped the spirit of the animal or object would become active in the supernatural realm and somehow trap or scare evil influences away.[1]

A familiar was a small animal thought to be inhabited by a malevolent spirit who would carry out a witch's wicked behests. Cats, ferrets, toads and even chickens were included in this superstitious list. In the infamous case of the Witches of Warboys, in Cambridgeshire, Alice Samuel, her husband and her daughter were all hanged for witchcraft between 1589 and 1593. Much of the testimony centred on Alice and her brown chicken. The Devil was said to have been sent by Alice 'in the likenes of a dun* chicken, & would talke familiarly with them, saying that they came from mother Samuel, (whom they called their dame) and were sent by her to the chyldren to torment and vexe them in that sort'.[2]

In the seventeenth-century East Anglian witch trials, led by England's most notorious witch-finder, Matthew Hopkins, over one hundred and twenty Suffolk women and men were charged with witchcraft and tried at Bury St Edmunds. Many

* From the Old English *dunn* meaning dull brown or dark coloured.

of the confessions extracted by Hopkins described possession by malevolent creatures, including a chicken named 'Nan' and other nameless backyard poultry. Based on the firm belief that these creatures were agents of an evil force, nearly seventy of the accused were executed.[3]

Throughout history, chickens have unwittingly attracted superstitious beliefs, both good and ill. The bird holds a strangely ambiguous position in the world of folk magic, at times a totem of good fortune, at others a tool of witchcraft or harbinger of doom. Eggshells invited a particularly rich array of superstitions. The first-century poet Persius mocked his fellow Romans' belief in the power of cracked eggshells to attract evil spirits, writing: 'Then a cracked egg shell fills you with affright, And ghosts and goblins haunt your sleepless night.'[4] The belief came from a method of divination that involved roasting eggs in an open fire; if the egg burst before it was fully cooked, bad luck would surely follow.

Pliny the Elder noted that 'as soon as anyone has sucked the juice of eggs, they are immediately broken'.[5] The belief that eggshells – once their contents were eaten – must be smashed or pierced continued well into the Middle Ages and beyond. Sixteenth-century French scholar Adrianus Turnebus warned that 'if the shell be perforated with a needle, witches will have the power of injuring the person who has eaten the meat, and therefore the prudent always crush the shells'. A century later, Thomas Browne documented the beliefs of the credulous in his brilliant *Vulgar Errors* (1646), explaining: 'To break the egg shell after the meat is out, we are taught in our childhood, and practise it all our lives; which neverthelesse is but a superstitious relict [...] and the intent hereof was to prevent witchcraft; for lest witches should draw or prick their names therein, and veneficiously mischiefe their persons, they broke the shell.'[6]

One of the most extraordinary beliefs was that a witch could use an empty eggshell as a boat for her devilish sorties. Reginald Scot's *The Discoverie of Witchcraft* (1584), a book that both argued against the existence of witchcraft and defended those who had been accused of being witches, recorded that the gullible believed witches could 'saile in an egge shell, a cockle or muscle shell, through and under the tempestuous seas'.[7] Witches, it was thought, could shrink themselves and climb aboard these tiny, eggy craft. Ben Jonson's *Masque of Blackness* (1605) took the idea even further, adding a pin for a mast and a spider's web for rigging:

A flash of Light, and a clap of Thunder,
A storm of Rain, another of Hail.
We all must home, i'the Egg-shell sail;
The Mast is made of a great Pin,
The takle of Cobweb, the sail as thin,
And if we go through and not fall in –

Although such ideas seem charming to a modern onlooker, these quaint notions were deadly. Despite the odd rational voice, such as Scot's, many people continued to hold on to medieval beliefs throughout the seventeenth, eighteenth and even nineteenth centuries. Educated and influential people of the day, including King James I and Members of Parliament, believed utterly and wholeheartedly in the existence of witches and their ability to 'reduce their size, sail in shells, and wage a very real war of chaos and evil with borrowed power from the microscopic world, if not the underworld'.[8]

Grace Sherwood was the last person known to have been convicted of witchcraft in Virginia, in 1706; one of her 'crimes', according to neighbours, was that she had sailed across the Atlantic to the Mediterranean in an eggshell. According to her accusers, on her arrival in Europe Grace dug up a rosemary plant, brought it back to American shores and planted it in her garden, where the herb soon acclimatised and spread. As one Victorian journalist later noted, 'It was true that it harmed nobody, but such proceedings were uncanny. Only witches who rode on broomsticks to midnight meetings could sail in egg-shells, and as witches cast spells and exerted malignant influences on honest people, it was plain that Grace Sherwood was a witch, and ought to be tried and punished.'[9]

Writing in the *Edinburgh Magazine* in 1818, one Scottish journalist commented on the persistence of the belief in some of the older members of the community: 'Here, *Noroway*[*] [Norway] is always talked of as the land to which witches repair for their unholy meetings. No old-fashioned person will omit to break an eggshell, if he sees one whole, lest it should serve to convey them thither.'[10] Knowledge of the eggshell superstition existed well into the twentieth century. After the First World

* Noroway was the old Scots dialect for Norway.

War, Rudyard Kipling penned 'The Egg-Shell', a poem that drew on the ancient trope,* while just a few years later Scottish poet Elizabeth Fleming turned the superstition into a well-known nursery rhyme:

Oh, never leave your eggshells unbroken in the cup
Think of us poor sailor-men and always smash them up,
For witches come and find them and sail away to sea,
And make a lot of misery for mariners like me.

Eggshells were also an important element in folk tales about changelings, supernatural beings who were thought to have been left in place of a human child or baby. References to changelings appear in England from the sixteenth century onwards – it was thought that fairies or elves would steal human babies and substitute one of their own kind, who was often elderly and bad-tempered. Superficially, the baby would look the same and so it was necessary for parents to establish whether their 'child' was an imposter by subjecting it to various trials such as leaving it by a hot fire or outside overnight. Abuse wasn't the only way that mothers and fathers could reveal that their child was in fact a changeling; another method was to try to force the imposter

* THE WIND took off with the sunset—
The fog came up with the tide,
When the Witch of the North took an Egg-shell
With a little Blue Devil inside.
"Sink," she said, "or swim," she said,
"It's all you will get from me.
And that is the finish of him!" she said,
And the Egg-shell went to sea.

to laugh or speak out loud by performing the surprising feat of cooking in an eggshell. The Grimm brothers, in their early nineteenth-century *Children's and Household Tales*, retell a time-honoured story from German folklore that sees the changeling blurting out his age and real character:

> 'A mother had her child taken from the cradle by elves. In its place they laid a changeling with a thick head and staring eyes who would do nothing but eat and drink. In distress she went to a neighbour and asked for advice. The neighbour told her to carry the changeling into the kitchen, set it on the hearth, make a fire, and boil water in two eggshells. That should make the changeling laugh, and if he laughs it will be all over with him. The woman did everything just as her neighbour said. When she placed the eggshells filled with water over the fire, the blockhead said: "Now I am as old, As the Wester Wood, But have never seen anyone cooking in shells!" And he began laughing about it. When he laughed, a band of little elves suddenly appeared. They brought the rightful child, set it on the hearth, and took the changeling away.'[11]

The same 'cure' for changelings was also found on British shores. In 1892, folklorist Joseph Jacobs compiled a book of Celtic folklore. In it, the same story appeared, only with twin babies and the location moved to Treneglwys, a village in Powys, Wales. Jacobs' retelling of the 'Brewery of Eggshells' recounts the mother making potage in an eggshell and pretending to give it to a band of reapers. Knowing that such a small meal could never feed an army of hungry farm workers, one of the changeling twins shouted to the other: 'Acorn before oak I knew, An egg before a hen, But I never heard of an eggshell brew, A dinner for harvest men.' With that, the mother seized the babies and threw them into the river, where they were rescued by elves who swapped them back for her real children.[12]

It's significant that a hen's eggshell is used in the ritual to expose the changeling. The long-held belief that eggshells could attract malevolent forces often encouraged people to create counter-charms, which would lure evil spirits out into the open where they could be captured. Some of the most numerous examples of these 'demon traps' come from archaeological sites in both Iraq and Iran and date from between the sixth and eighth centuries. Also known as 'incantation bowls', thousands of these ancient counter-charms have been discovered over the years; they consist of an upturned earthenware bowl under which an inscribed eggshell had been placed. The bowls were then buried in the hope that they would lure and disable any malevolent spirits, especially those that caused harm to people and their property.[13] And while the majority of these magical finds date to the early Middle Ages, the practice had been known since Roman times. In AD 17, for example, the town of Sardis, in what is now western Turkey, was destroyed by a vast tremor; when the inhabitants rebuilt their homes, they placed demon bowls and eggshells under the buildings' foundations to ensnare any remaining earthquake-causing spirits.

While empty shells presented both problems and opportunities for the superstitious, the complete, unbroken egg has been one of the most important metaphors in world religion. For many cultures, including some of our most ancient civilisations, the egg represented the birth of the world or universe – the idea of potentiality, of nothing becoming something. A chick hatching, after a long period of dormancy, was also a powerful emblem for rebirth and renewal. In ancient Greek mythology, the first deity was said to have emerged from a cosmic silver egg. This 'Orphic egg', which was laid by Chronos, the personification of time, opened to reveal Phanes, creator of all other Greek gods.

Versions of this 'cosmic egg' cosmology feature throughout world belief systems and philosophies. The egg often contains a divine being who hatches out and goes on to create the Earth, its people and the heavens. Pangu, the creation figure in Chinese mythology, was said to have emerged from a cosmic egg; the Dogon of Africa tell a similar tale, of a cosmic egg containing two sets of twins; and in Tahitian myth Ta'aroa, the supreme deity and creator, broke out of a cosmic egg and used the shell and his own body to form parts of the world. Early Egyptian, Japanese, Bantu, Slavic, Hindu, Finnish and other seemingly disparate cultures have surprisingly similar themes.

And while Christianity makes no mention of cosmic eggs in its teachings, the motif is hiding in plain sight. Writing in the middle of the eighteenth century, French scholar Antoine

Court de Gébelin mused that the custom of giving eggs at Easter can be traced back to these early theologies of 'the Egyptians, Persians, Gauls, Greeks, Romans, &c. among all of whom an Egg was an emblem of the universe, the work of the supreme Divinity'.[14] Christians, he argued, adopted the egg, and the chick, as symbols of the Resurrection because they were such simple and potent metaphors for Jesus rising from the dead after being entombed.

The hen's egg also plays an important part in the Easter festivities. Eggs were one of the foods traditionally forbidden during Lent, the period of fasting in the run-up to Easter. Its origins are lost in time but the forty-day abstinence was widely practised among Christians by the fourth century. No alcohol, meat or animal products were allowed during these few weeks, as a way of commemorating Jesus' forty days in the desert. Any eggs that were laid during Lent were hard-boiled and kept for Easter Sunday, when normal eating could return with gusto. The tradition of decorating these boiled eggs soon followed and may have its roots in early Orthodox Christianity, the church that developed out of the Eastern Roman Empire. Followers would stain eggs red as a reminder of the blood spilled by Christ during his Crucifixion, a practice that's still common among Greek, Middle Eastern, Russian and Slavic cultures.

One of the most cheerful descriptions of this venerable tradition was recorded in the sixteenth century by the writer Richard Hakluyt in his remarkable *The Principall Navigations, Voiages and Discoveries of the English Nation*, a vast collection of real-life explorers' tales from the furthest corners of the known world. Of the Russians, he wrote, 'They have an order at Easter, which they alwaies observe, and that is this: every yeere, against Easter, to die or colour red, with Brazzel [Brazil wood], a great number of Egges, of which every man and woman giveth one unto the priest of the parish upon Easter Day in the morning.' Part of the joy of the celebration involved swapping coloured eggs as a token of affection. 'For when two friends meete during the Easter Holydayes they come and take one another by the hand; the one of them saith, "The Lord, or Christ, is risen"; the other answereth, "It is so of a trueth"; and then they kiss, and exchange their Egges, both men and women, continuing in kissing four dayes together.'[15]

The exchange of decorated eggs continues to be an important part of Russian Easter celebrations. Perhaps the most illustrious, however, were those given by the Romanov Imperial family during the late nineteenth and early twentieth centuries. As one Victorian traveller noted with amazement: 'Scarcely any material is to be named that is not made into Easter eggs. At the Imperial glass cutting manufactory we saw two halls filled with workmen employed on nothing else but in cutting flowers and figures on eggs of crystal. Part of them were for the emperor and empress to give away to courtiers.'[16]

Most famously of all, the Tsar's family presented each other with eggs designed by Peter Carl Fabergé, jeweller to the court. His exquisite creations transformed the humble folk tradition of gifting eggs into an opulent art form. The 'First Hen' Fabergé egg was gifted in 1885, a present from Tsar Alexander III to his wife to take her mind off the political turmoil outside the palace walls. Made from gold, coated in white enamel, the egg contained a golden yolk, which itself concealed a tiny gold hen with ruby eyes. The hen then opened to reveal two more surprises – a miniature gold and diamond crown and a minuscule ruby pendant. The gift was a triumph and became the first in a series of fifty Imperial Fabergé Easter eggs presented between 1885 and 1916. When hunger and failed harvests ravaged the empire at the beginning of the twentieth century, however, the Fabergé eggs began to epitomise a ruling elite painfully out of touch with its people. After the Bolshevik Revolution in 1917, some of the confiscated eggs were sold, while others disappeared into the ether. Of the fifty Imperial eggs made by Fabergé, forty-three are now in private collections or museums but seven remain missing,* the last undiscovered treasures in the world's most expensive Easter egg hunt.

The seasonal gifting of decorative eggs has also been a tradition in England since the Middle Ages. The earliest documented mention of Easter eggs comes from the household accounts of King Edward I in 1290, when he ordered 450 boiled eggs, covered in gold leaf, to be presented to the royal household.[17] Edward called them 'pace eggs', a term used across parts of northern England well into the twentieth century. 'Pace' was a corruption of the Latin *pascha*, itself a derivative of

* It was eight eggs, until recently, when an American scrap-metal dealer speculatively bought a gaudy gold ornament from a flea-market stall in the hope he could turn a quick profit by melting it down. A quick internet search of the egg's markings revealed it to be the Third Imperial Easter Egg, made in 1887, two years after the 'First Hen' egg, and subsequently sold in 2014 for an undisclosed amount believed to be over £20 million.

the ancient Hebrew *pesah*, or Passover. In Lancashire, Yorkshire and southern areas of the Lake District, poor villagers would dress up for 'pace-egging', collecting food for an Easter feast by begging door to door. In the rite, which was not unlike trick-or-treating, pace eggers would dress up or blacken their faces and recite a rhyme, in the hope that the householder might part with a small bit of bacon, a sliver of cheese or a hard-boiled egg:

We are two-three jolly boys, all of one mind,
We are come a-pace egging, and we hope you'll prove kind.
We hope you'll prove kind with your eggs and strong beer,
And we'll come no more a-pace-egging until another year.[18]

Mendicancy, however, had been illegal in England since the 1531 Vagabonds Act, legislation that prevented the able-bodied poor from begging. To circumvent the rules, pace eggers began to put on dramatic performances in return for donations.[19] Pace Egg plays were often retellings of medieval mystery plays or biblical tales, with good-natured dances, mock battles and bawdy songs centred around the themes of death and resurrection, with comic stock characters such as The Doctor, Owd Bett and the fantastically named drunkard, Old Tosspot.*

Hens' eggs and Easter have been intimately intertwined for centuries, but one tradition is perhaps not as old as you might

* 'Tosspot' is a colourful medieval insult for a drunkard. It means someone who has been literally and copiously tossing back the booze, as in 'If any poore man have in a whole week earned a grote, He shal spend it in one houre in tossing the pot' by sixteenth-century playwright Ulpian Fulwell.

imagine. A common misconception is that the Easter Bunny derives from celebrations of the ancient pagan goddess of spring, Ēostre, and her loyal hare companion, who together symbolised fertility and renewal. Much of this speculation is based on neo-pagan fantasy. The learned monk Bede, in the eighth century, documented the different names that Anglo-Saxon tribes gave to their months before the arrival of Christianity. April, he claimed, had once been known as 'Ēosturmōnaþ' or Ēostre's month, and was a time of feasting in her honour. And while scholars believe there may be some truth in Bede's description of this pagan goddess, he never actually made any mention of hares, eggs or any other recognisable Easter traditions.

The earliest solid evidence for the egg-bearing rabbit comes, in fact, much later. In 1682, German physician and botanist Georg Franck von Franckenau described a Germanic folk tradition of Easter egg hunts: 'they call these eggs *di hasen-eier* [hare's-eggs] from the story, told by children and the simple-minded, that a hare (*der Oster-Hase*) [the Easter-Hare] laid the eggs to hatch hidden in the garden's grass, bushes, etc., where they are eagerly sought out by the children to the delight of the smiling adults.'[20] The custom of the Easter Bunny and coloured hens' eggs was then taken to America by emigrants from south-western Germany in the eighteenth century, where it soon became a household tradition. The Easter Bunny 'arrived' on British shores a few decades later, thanks to Queen Victoria's mother, the German-born Duchess of Kent. In 1833, a teenage Victoria wrote affectionately about an egg hunt arranged at Kensington Palace: 'Mama did some pretty painted &

ornamented eggs, & we looked for them.'[21] After she met and married Albert, Queen Victoria continued the tradition, to her children's delight, hiding hens' eggs in 'little moss baskets' for them to find on Maundy Thursday.[22]

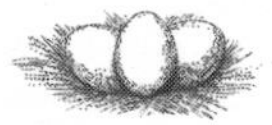

The egg's association with creation myths and fertility also made it ripe for love divinations, spells thought to attract a partner or make someone fall in love with you. Deep in the bowels of the British Library sits a collection of seventeenth-century medical and kabbalistic prescriptions, including a delightful Hebrew love potion and marital aid: 'For love between husband and his wife or even if he is under a spell [i.e. impotent]: take spring water, wine, and myrrh, and pepper, and two dove eggs and two hen eggs and break them, and mix them together, and give the mixture to drink to the man and the woman, and they will love each other.'[23]

Records of English folklore also talk about 'oomancy', a form of romantic fortune-telling using eggs. In a practice known as early as the seventeenth century, young women and men would drop egg white into water and interpret the shapes made by the swirling albumen. Different patterns would indicate their future partner's occupation or personality. Similar rituals were seen in Denmark in the nineteenth century according to Edwin and Mona Radford, early twentieth-century collectors of European folklore: 'There, on New Year's Eve, members of both sexes take a newly laid egg, perforate with a pin the smaller end and let three drops of the white of the egg fall into a basin or bowl of water. The drops diffuse themselves over the water into fantastic shapes of what look like trees. From these, so it averred, the fortune of the egg dropper, the character of his wife-to-be (or husband-to-be) and the number of children from the marriage can be told!'[24]

The chosen days for the rituals were often significant – New Year's Day, Midsummer Day or, more often, St Agnes' Eve.* The last of these – which falls on 20 January – was traditionally the night when unmarried girls performed another ritual, which involved eating an eggshell laced with salt and walking to bed backwards, hoping they would dream of their future husbands. John Keats even wrote about the tradition in his poem 'The Eve of St Agnes':

They told her how, upon St Agnes' Eve,
Young virgins might have visions of delight,
And soft adorings from their loves receive
Upon the honey'd middle of the night.[25]

The association of romantic love and hens' eggs may also have a connection to the chicken's seasonal laying habits. Commercial

* St Agnes was a virgin martyr and the patron saint of chastity and young women. According to tradition, Agnes was a beautiful girl born into Roman nobility and pursued by many suitors. One man, slighted by her refusal, succeeded in having Agnes charged with the crime of being a Christian and sentenced to death by burning. The young martyr was tied to the stake but the wood would not light and the guards were forced to behead her. She was thought to be only twelve or thirteen when she died.

egg chickens now lay all year round but, traditionally, hens began laying in spring after a long break over winter. Valentine's Day was often viewed as *the* day when the best hens would start to lay, with slower, less productive birds following suit over the next fortnight. By 1 March, St David's Day, or 2 March, St Chad's Day, all hens should be back to laying, as one almanac of folklore recorded: 'By Valentine's Day, Every good hen, duck or goose should lay, By David and Chad, Every hen, duck or goose should lay, good or bad.'[26]

While eggs often represented female fertility, the cockerel was synonymous with male potency in medieval life. An earthy song from 1670, 'The Young Gallants Tutor, Or, An Invitation to Mirth', mentions a boozy selection of favourite tipples: 'With love and good liquor our hearts we do cheer, Canary and Claret, Cock Ale and March beer.' Beer, claret and canary,* a sherry-like wine, are recognisable forms of drink today, but 'Cock Ale' has thankfully slipped into obscurity. One of the oldest versions of the recipes was written by Sir Kenelm Digby. Digby was a prominent English courtier and diplomat but also penned a fascinating cookbook, *The Closet of the Eminently Learned Sir Kenelme Digbie Kt. Opened*, which was published posthumously in 1669 by his faithful servant. His instructions for the cockerel-brew were as follows:

* 'Canary' or 'canary sack' is an almost obsolete sweet, strong fortified wine imported from the Canary Islands, which was incredibly popular in the sixteenth and seventeenth centuries. Ben Jonson's poem 'Inviting a Friend to Supper' (1616) mentions 'a pure cup of rich Canary wine, Which is the Mermaid's now, but shall be mine'. The closest equivalent today is a sweet, aged wine called Malvasía Canari, made in Lanzarote.

> 'To Make Cock-Ale. Take eight gallons of Ale; then take a March cock and boil him well; and take four pounds of raisins well stoned, two or three nutmegs, three or four flakes of mace and half a pound of dates. Beat all these in a mortar, and put to them two quarts of the best sherry-sack. Put this into the Ale, with the Cock, and stop it close six or seven days, and then bottle it: and after a month you may drink it.'

This spiced, sweetly alcoholic chicken broth doesn't sound like something you could down by the pint but English men, and women, swore by its potency. Nathan Bailey's *Dictionarium Britannicum* (1736) described it as a 'pleasant drink, said to be provocative', while a jolly Victorian book of slang noted that it was a 'homely aphrodisiac'.[27] Men in the late seventeenth century apparently needed it more than most. 'The Women's Petition Against Coffee', a pamphlet issued in London in 1674, made the shocking and controversial claim that coffee was making the men of the capital impotent in bed. The pamphlet, which was probably making a political point about dissenters meeting in coffee houses, disguised as bawdy satire, called for a ban on coffee-drinking among men below the age of sixty and instead encouraged the drinking of 'Cock-Ale' to spice things up in the bedroom, lest London's men 'run the hazard of being Cuckol'd by Dildo's'.[28]

Male sexuality and the symbolism of the cockerel were, and still are, deeply entangled. Writing in the early nineteenth century, the naturalist Georges-Louis Leclerc, whom we met in Chapter 1, commented that any cockerels seen engaging in same-sex encounters had historically been given short shrift, noting that a 'law mentioned by Plutarch, in which it was enacted, that a Cock convicted of this unnatural act, should be burnt alive'.[29] Hens who behaved like cockerels were also viewed as trouble and a metaphor for the dangers of allowing women a voice: 'If your hen chance to crow,' advised the seventeenth-century cookery writer Gervase Markham, 'which is an evill and unnatural infirmity in her, you shall forthwith pull her wings and give her wheat scorched and mixt with powder of Chaulke'.[30] The sentiment continued in traditional sayings well into the twentieth century, including the British 'a whistling woman and a crowing hen are neither fit for God nor men', and its American variation 'a whistling woman and a crowing hen never come to a very good end'.

In 1474, a cockerel was put on trial and prosecuted in the city of Basel, Switzerland; the creature was charged 'for the heinous and unnatural crime of laying an egg' and condemned to be burned at the stake. The act of a male bird laying an egg was so aberrant that the bird's execution was greeted 'with as great solemnity as would have been observed in consigning a heretic to the flames, and was witnessed by an immense crowd of townsmen and peasants'.[31] The residents believed they were right to be worried - eggs laid by 'cockerels' were thought to be the instruments of witchcraft and used in terrible spells. Moreover, if allowed to hatch, the eggs would contain cockatrices or basilisks, mythical and terrifying beasts with the head of a cockerel and the tail of a serpent, which could kill all living things with just a glance or puff of breath. Anyone lucky enough to capture and kill one, however, could use its ashes to make gold. The twelfth-century Benedictine monk Theophilus Presbyter wrote with great confidence that Arab alchemists bred these fearsome creatures by trapping two aged male roosters and overfeeding them until they mated and lay basilisk chicks.[32] His recipe for gold required a heady concoction of powdered basilisk chick, human blood, copper and vinegar.

The idea of a cockerel laying an egg must have been mystifying to the medieval onlooker, but modern science has shown that hens can develop cockerel-like characteristics. Certain medical conditions that damage a hen's ovary can cause the bird to start producing androgens, hormones that are largely responsible for the behaviour and characteristics of the male chicken. The female bird will start to develop male plumage, a wattle and even crow like a cockerel but still be capable of laying eggs.

Sex expression in poultry may explain the Basel chicken trial, but 'cock's egg' superstitions also arose from the common quirk that is yolkless eggs. When a young hen or 'pullet' starts to lay eggs at around six months of age, her first attempts can often be small and yolkless. These incomplete, sterile eggs could never hatch a chick and so were thought to have been laid by cockerels; backyard poultry keepers still call these fairy eggs, fart eggs or witch eggs. Far from being a long-forgotten superstition, some cultures still hold this belief. Social anthropologists were recording similar views among the Mambila people of Cameroon in the late twentieth century: 'Oh yes, cockerels lay eggs, but small ones,' explained one villager. 'You can eat them if you like, but what you should do is to weave a small basket, put the egg in it and then hang the basket at a crossroads. Then your chickens will grow well and fat and not die and they will lay many eggs.'[33]

If a cockerel crowed at the 'incorrect' hour or unexpectedly, it didn't bode well. In *The Satyricon*, a first-century work of fiction attributed to the Roman courtier Gaius Petronius Arbiter, one of the characters panics on hearing a cock crowing: 'He had not ceased speaking when a cock crowed! Alarmed at this omen, Trimalchio ordered wine thrown under the table and told them to sprinkle the lamps with it; and he even went so far as to change his ring from his left hand to his right. "That trumpeter did not sound off without a reason," he remarked; "there's either a fire in the neighbourhood, or else someone's going to give up the ghost. I hope it's none of us!'[34]

Three hundred years later, St John Chrysostom, early church Father and archbishop of Constantinople, despaired at people's continuing belief in natural omens: 'the Greeks are always children', he fumed at the end of the fourth century, 'spending much zeal in the pursuit of riches, and yet supposing the whole is undone by the crowing of a single cock'.[35] Throughout the medieval period, the cockerel's crow was carefully scrutinised. Cock-a-doodle-doodling at the wrong time, especially during the night, was an augury of an untimely death, a superstition that continued well into the nineteenth century. In Thomas

Hardy's *Tess of the D'Urbervilles*, published in 1891, the death of the eponymous heroine is forewarned by a cockerel's crow on her wedding day:

> 'The white one with the rose comb had come and settled on the palings in front of the house, within a few yards of them, and his notes thrilled their ears through, dwindling away like echoes down a valley of rocks. "Oh?" said Mrs Crick. "An afternoon crow!" Two men were standing by the yard gate, holding it open. "That's bad," one murmured to the other.'

A cockerel's poorly timed call might have sent shivers down the backs of our ancestors but the avian alarm clock, screeched at the correct hour, has long been viewed as auspicious. Zoroastrianism is the world's oldest continuously practised monotheistic religion and thought to have influenced the teachings of Judaism, Christianity and Islam. Its followers also believed in the sanctity of the cockerel. The idea of duality is central to a faith that stretches back far into Indo-Iranian prehistory – the idea that good and evil exist both in the universe and within the human soul, a struggle between two opposing forces that is endlessly played out.

In the Avesta, the central religious text of Zoroastrianism, the cockerel embodies the virtues of hard work and alertness. He is 'The herald of dawn. Among the domestic birds, the cock is the most sacred. As the admonisher of mankind to discard sloth, and to wake up early to lead an industrious life, he is the ally of the ever-wakeful Sraosha.'[36] The bird is also central to the well-being of the family unit; the Zoroastrian ideal was 'of a householder living with his wife and children, in a house in which cattle thrive, the fire grows, virtue increases, welfare arises, and the dog prospers. The cock may be added to the possession of a happy home, for it is the duty of this domestic bird to wake the members of the family at dawn.'[37]

In Chinese mythology, the cockerel was also a creature who embodied noble attributes. The 'Bird of Five Virtues', the cockerel represented wisdom, ferocity, courage, humanity and faithfulness; its red comb was thought to resemble a government official or scholar's hat, hence its association with intelligence; its spurs linked the bird to martial skill and bravery. The cockerel's diligent protection of its hens, and willingness to share food, represented civility, while its regular dawn chorus showed the world the importance of reliability. Interestingly, the Chinese character for cockerel (鸡 jī) also has a similar pronunciation to the symbol for 'lucky' (吉 jí), leading many to view the bird as an emblem for good fortune.

In the Bible, the cockerel has come to hold a number of meanings. In the New Testament, during the Last Supper with his disciples, Jesus predicted that Peter would deny all knowledge of him before the cockerel crowed the following morning. According to the Gospel of Matthew, Peter insisted, 'Even if all fall away on account of you, I never will.' 'I tell you the truth,' Jesus replied, 'this very night, before the rooster crows, you will disown me three times.' But Peter declared, 'Even if I have to die with you, I will never disown you.' The Bible tells us that, after

Jesus' arrest, Peter did indeed deny knowing him three times but on the last denial heard the cockerel crow and, realising the truth, began to weep bitterly.

After the Resurrection, the Christian story sees Peter reaffirming his love for Jesus and going on to become one of the early leaders of the church, the first pope and a saint. The cockerel became St Peter's symbol – a representation of Christ's power to forgive sins, put flight to the powers of darkness and start afresh. Pope Nicholas I, the bishop of Rome from 858 AD until his death in 867 AD, decreed that a figure of a cockerel should be placed on every church as a reminder of Peter's journey. Many steeples are still adorned with chicken weathervanes to this day.

The oldest surviving weathercock comes from the bell tower of the Church of Saints Faustino and Giovita in Brescia, Italy. Dating from the ninth century, it guarded the bell tower for over a thousand years until the late nineteenth century, when it was removed and placed in the town's museum for safekeeping. One of the most iconic weathervanes, however, sat atop Notre-Dame's spire in Paris. The hollow copper cockerel was said to contain three religious relics – a remnant from the Crown of Thorns and remains of Saints Denis and Genevieve, both patron saints of Paris. Amazingly, even when the 300-foot wooden spire toppled in the catastrophic fire of 2019, the metal bird miraculously landed on the pavement and survived, dented but intact, no doubt a potent sign to the faithful.

The connection between the cockerel and the new day inevitably led to religious associations with light and spiritual awakening. Hugo of Saint Victor, a twelfth-century theologian, explained in his *Mystical Mirrour of the Church* that: 'The sleepers be the children of this world, lying in sins. The cock is the company of preachers, which do preach sharply, do stir up the sleepers to cast away the works of darkness, which also do foretell the coming of the light.'[38] The power of the cockerel's crow to vanquish dark forces was also deeply embedded in folklore. According to English local legend, when early Christian churches began springing up across the county of Sussex, the Devil decided to dig a huge ditch to let in sea water and flood the residents. As he was digging, throwing great sods of earth behind him, he created many of the local landmarks, including Rackham Hill and Chanctonbury Ring. An old woman managed to scare the Devil away by lighting a candle and forcing her cockerel to crow. Fooled into believing that it was daybreak, the Devil fled, leaving the ditch only half finished.[39]

In the late nineteenth century, leading Celtic scholar Sir John Rhŷs recorded traditional Welsh and Manx folk beliefs in faeries – mischievous and unpredictable otherworldly spirits whose night-time antics could be dispelled by the crowing of a cockerel: 'At midnight to the minute,' he wrote, 'they might be seen rising out of the ground in every combe and valley; then, joining hands, they would form into circles, and begin to sing and dance with might and main until the cock crew, when they would vanish.'[40] Audiences watching Shakespeare's first performances of *Hamlet* would have understood why the ghost disappeared at hearing the cockerel's cry. The conversation between Bernardo, Horatio and Marcellus revealed the power of the bird to drive away the supernatural:

Bern:

It was about to speak, when the cock crew!

Hor:

And then it started like a guilty thing
Upon a fearful summons. I have heard
The cock, that is the trumpet to the morn,
Doth with his lofty and shrill-sounding throat
Awake the god of day.[41]

Not all the cockerel's qualities were admired, however. One person's pride is another's self-importance, and the bird's strutting, confident ways made it ripe for allegory. The best-known of these tales began life as *Roman de Renart*, a twelfth-century fable about a fox and cockerel. In the story, Renart the fox sneaks into a chicken enclosure, looking for a meal. The hens scatter but the cockerel, Chanticleer, whose bravado outweighs his brains, stands his ground. Renart decides to use flattery to trick Chanticleer into showing off his singing voice. As soon as Chanticleer closes his eyes and stretches his neck to crow, Renart pounces. With the cockerel's neck in his mouth, the fox makes a dash for it, with the farmer in hot pursuit. Chanticleer, using the same ruse as Renart, dares the fox to shout insults at the man giving chase. Unable to resist the chance to boast, Renart opens his mouth, only for Chanticleer to make his escape.

The moral of the story was clear – beware false flattery as pride comes before a fall – a motif that continued to chime with audiences throughout subsequent reworkings of *Roman de Renart*, including Chaucer's 'The Nun's Priest's Tale', written at the end of the fourteenth century, and John Dryden's 'The Cock and the Fox' published in 1700, the year he died. For Dryden, a wonderful satirist and lampooner of court politics, the lesson of the cockerel clearly struck a chord:

In this plain fable you the effect may see
Of negligence, and fond credulity:
And learn besides of flatterers to beware,
Then most pernicious when they speak too fair.[42]

'Cock-throwing' or 'cock-shying' was a hugely popular pastime among the English, especially on Shrove Tuesday. As one anonymous journalist described it in his 1737 *Enquiry into the Original Meaning of Cock-Throwing on Shrove-Tuesday*: 'Battering with missive Weapons a Cock tied to a Stake, is an annual Diversion that for Time immemorial has prevailed in this Island [and] peculiar to our Nation.'[43] The rules of the cruel game were simple – a hapless cockerel would be tethered to a post by a length of rope and people would take turns throwing stones or weighted wooden sticks called 'coksteles' at it. Participants might part with a few coins for three throws and whoever dealt the fatal blow, won. Or, if you managed to knock the poor bird over, the cockerel was yours if you could grab it before it scrambled back to its feet.

The choice of animal was interesting, however, and may have been linked to England's long-standing friction with France. The two nations had been at loggerheads for centuries. Edward III, king of England from 1327 to 1377, had led England into the Hundred Years War, a long struggle between the two countries over the succession to the French throne, and the nations continued to spar throughout the reigns of subsequent monarchs. The cockerel had the misfortune to have the same Latin name – *gallus* – as the inhabitants of Gaul, *Gallus*, 'so that nothing could so well represent, or be represented by the One as the Other'. 'The Frenchman is ingeniously Ridicul'd and Bastinado'd in the Person of his Namesake. This naturally accounts for the cruel and barbarous Treatment poor Chanticleer has undeservedly met with.'[44]

While the link between the cockerel and France may have induced some English folk to pick up their coksteles, it seems that collective celebrations involving barbarity towards chickens may have had even older roots. The two 'sports' allowed on Shrove Tuesday – tormenting cockerels and playing football – were recorded as early as the twelfth century by William Fitzstephen, cleric and administrator to Thomas Becket, the soon-to-be-martyred archbishop of Canterbury. Repeated attempts down the centuries to prohibit these rowdy pastimes suggest that raucous events were a common part of the Lent season. Thomas Crosfield, in the early 1600s, recorded a continuation of these customs – 'throwing at cockes' and 'footeball' were all part of a day of general 'frittering' by the masses.[45]

Even boys' schools participated in these unruly pursuits. William Henderson, who compiled folklore traditions in northern England at the end of the nineteenth century, described the annual event: 'Foot-ball and cock-fighting were the great diversions on what was called Fastens Eve* [...] My father-in-law used often to speak of the cock-fights which regularly took place in all schools on that day. The master found the cocks, but the boys paid for them. There was a regular subscription for the purpose, each boy giving what was called a "cock-penny". The masters made a good profit out of the transaction, as they were entitled besides to claim all the runaway birds, which were called "Fugees" [...] I learn from a clergyman, formerly a scholar at the grammar-school of Sedbergh, in Yorkshire, that the master used to be entitled to 4½d. yearly from every boy on Shrove Tuesday to buy a fighting-cock.'[46]

Cock-throwing and cockfighting, it seems, were just two out of a whole list of games played at the chicken's expense. Cock- or hen-threshing, which was also called 'whip the cock', was particularly popular. *The Youth's Cornucopia*, a children's book from 1832, described the game thus: 'The custom was to tie a poor hen at the back of some bumpkin or other and decorate him with small bells, and then his companions, who were blind-folded, threshed him and the hen together as well as they could with boughs, guided by the sound of the bells: so that, as they had a better clue to his position than in the case of blindman's-buff, it must have been a very lively scene,

* *The evening of Shrove Tuesday.*

and sufficiently amusing, but for the thoughts of the torture the poor bird endured.' After the 'fun' of the evening, the poor bird would apparently be boiled with bacon and served with pancakes and fritters.[47]

While the English mocked the French cockerel and punished it for pleasure, *le coq gaulois* was gradually adopted as a proud national emblem in its own country. Louis XIV employed the bird as one of a handful of royal symbols along with the *fleur-de-lis*, the crown and the radiant sun, each with its own symbolic meaning. The powerful, dawn-crowing cockerel served as an effective visual device for a man who threw himself into absolute rule, military glory and daily religious observance. Ironically, the cockerel also later became one of the motifs of the French Revolution. Propaganda posters and literature from the late eighteenth century combined the slogan *Liberté, Egalité, Fraternité* with the trinity of republican motifs: the tricolour flag, a deliberately simple design to contrast with the extravagance of royal banners of the *Ancien Régime*; the Phrygian cap, a hat that had once symbolised freed Roman slaves; and the cockerel, the bird that represented both eternal vigilance and the long rural heritage of the French people.

During the First World War, the Gallic *coq* perfectly embodied France's resistance and courage in the face of enemy troops, a peasant's animal that showed pride and bravery in combat with the Prussian eagle. Postcards, recruitment posters and political cartoons all celebrated the bird's pugnacity and spirit but nothing more embodied the *coq gaulois* than the deeply patriotic war song of 1918, '*Cocorico! Ou l'aigle et le coq*' ('Cock-a-doodle-doo! Or the Eagle and the Cockerel'). Hundreds of years of imagery and symbolism of the cockerel came together to celebrate France's triumph: the ferocity of the ancient fighting fowl; religion's watchful, diligent bird; the idea of new beginnings and liberty; and the unapologetic victor all rolled into one:

And when he has flapped his wings,
Under freedom's sun,
The eagle understood one word, just one word,
Crowed across our land...
Brave lads! Brave lads!
Cock-a-doodle Doo!

4

METAPHORS

Pecking Orders, Cocks and Hot Chicks

Polish Chicken

Everyday speech is sprinkled with poultry metaphors. From pecking orders to broodiness, the bird and its behaviours have proved an almost inexhaustible source of inspiration. The cockerel, with its perceived libido and self-assurance, has become a byword for machismo; the hen, a symbol of everything loved and loathed about the female stereotype. The chicken provides handy similes when we cannot find the words ourselves - we talk of walking on eggshells, ruffled feathers, coming home to roost or having our wings clipped. People, like eggs, are good or bad. We mustn't count our chickens before they hatch. Stress makes us feel cooped up or like a headless chicken. Others, we moan, rule the roost or leave us henpecked.

Much of the language surrounding the chicken seems, at first glance, modern. And yet if we delve a little deeper, we find words and phrases that have been used for hundreds, if not thousands of years. Take 'chick', a word that means chicken but also, informally, an attractive young woman. To find how the word has changed, we need to go far, far back in time. The word 'chicken' was already known in Anglo-Saxon times - in Old English, the earliest form of the English language, chicken was *ćīcen*, pronounced something along the lines of 'cheeken', and referred to the young of the domestic bird. By the fourteenth century, 'chiken'* could mean the bird at any age, while the shortened version – 'chike' – was a baby or young chicken but also, sweetly, a child.

Some of the earliest examples use it in a disparaging way – a 'fendes chike' (fiend's child) or a 'deuels chyke' (devil's child) – but by the time of Shakespeare, in the early seventeenth century, the term was one of endearment for both boys and girls. In *The Tempest*, Prospero tells Ariel, an airy spirit often interpreted as a male character, 'My Ariel, chick, That is thy charge. Then to the elements Be free, and fare thou well!' Poets and writers loved to alliterate 'child' and 'chick', using both words to mean the same thing: 'Hee hath nor child nor chick to care for', imagined Renaissance dramatist Thomas Dekker in 1610. Britain's greatest polymath, William Morris, used the same analogy two hundred and fifty years later in his epic poem 'The Earthly Paradise' – 'But had no chick or child to bless this house'.[1] Indeed, the phrase to have 'neither chick nor child' had become an idiom of its own, meaning to be childless. In Bram Stoker's gothic horror *Dracula*, Mina Harker receives an inheritance from Mr Hawkins, who declares, 'I have left to me neither chick nor child. All are gone, and in my will I have left you everything.'[2]

It's often said that 'chick' got its sexualised meaning in the first decades of the twentieth century; its first appearance in text is regularly cited as *Elmer Gantry*, Sinclair Lewis's satirical novel about evangelical religion in 1920s America in which a woman

* Also spelled in a myriad of other ways – chikene, chyken, chykon and so on.

is described as a 'brainless little fluffy chick', but the word was probably already part of African American vernacular to describe a sexually attractive young woman, as was the similarly demeaning 'bantam'.[3] The word 'chick' became common parlance in the 1940s jazz scene, which popular culture of the 1950s and 1960s appropriated for its own. Elvis crooned that 'cats were born to give chicks fever', while Kerouac conjured up 'a beautiful young black chick' in his novel *On the Road*.

While some viewed 'chick' as a harmless compliment, by the 1970s an increasing number of women felt that the infantilising undertones of the word revealed its hidden misogyny. The queasy combination that 'chick' embodied - of both child-like vulnerability and sexuality - was problematic, to say the least, but the word proved annoyingly persistent. Even modern phrases - chick-flick and chick-lit - are used to suggest that women's cultural interests are vapid and childish, as fluffy and unthreatening as the mother hen's offspring. If you dig a little deeper, however, the word 'chick' to mean a desirable woman may be even older than Sinclair Lewis or its African American use. The phrase 'chick-a-biddy' was already well established by the late eighteenth century. Francis Grose's unputdownable *Classical Dictionary of the Vulgar Tongue* is a dictionary of slang words, first published in 1785. Grose, who collected colourful language from the sordid alleys of Georgian society, uncovered a wealth of descriptions of comely 'young lasses'. These included - alongside *goer, article, whip slang, bitch booby, dimber* and *fubsey*, the nickname *chick-a-biddy*, which meant a chicken but also a 'young wench'.[4]

In fact, many of the words we use to describe female chickens are loaded with double meaning. In Scotland, 'hen' is a wonderfully gentle term of endearment or familiarity for a girl or woman. Throughout historical Scottish literature, women are described as 'bonnie hens', 'thrifty hens' and 'wee hens', but the word – across much of northern Europe – actually started life as the term for a cockerel. Etymologists believe that the language spoken across Europe around 6,000–4,000 years ago had the specific word *kan*, meaning to sing or make a sound. From *kan* developed *hano* in very early Germanic, becoming *hana* in Old English, meaning cockerel, the bird who sings at sunrise. Unlike modern English, which doesn't have gendered nouns, Old English used masculine and feminine nouns, a practice that only disappeared in the twelfth century. The feminine version of a male chicken, *hana*, was *henn*. The male origin of the word 'hen' can still be seen in the modern usage of *Hahn*, the German word for cockerel (and in Norwegian *hane* and Dutch *haan*).

Although 'hen' is used largely as a term of affection in Scotland, across the rest of the British Isles it was given many of the sexist attributes historically associated with women. *Henwile,* a word meaning a sly but often rather inept trick, was used from the beginning of the seventeenth century. A *henwyfe* was a bawd or female brothel owner; a *hennie* an effeminate man or one who concerned himself with traditionally female matters. 'Hen-brained' meant scatty; a 'hen-house' a disparaging term for a female-dominated home.

The best-known term, however, is 'hen-peck', meaning to criticise and browbeat a subordinate husband. The hen-pecked man has an ancient pedigree and was a stock character of many a medieval comedy by the likes of Geoffrey Chaucer to John Lydgate, although the term was never expressly used. Women who subverted marital relationships through violence, scolding or cunning were often known as 'shrews', as in Shakespeare's *Taming of the Shrew*, but the role of the browbeaten husband had yet to be named.

By the seventeenth century, however, the unmanly 'hen-pecked' husband had become a common term, especially in the world of satire. Samuel Butler, popular poet and wit, delighted audiences with his sketches based on humorous characters he plucked from society; alongside 'the degenerate nobleman', 'melancholy man' and 'debauched man', his description of the 'henpect man' is one of the first literary uses of the phrase. 'The henpect man rides behind his wife', he wrote, 'and lets her wear the spurs and govern the reins. He is a kind of preposterous animal, that being curbed in goes with his tail forwards. He is subordinate and ministerial to his wife, who commands in chief, and he dares do nothing without her order.'[5]

At the turn of the eighteenth century, England's first Poet Laureate, John Dryden, wrote about a 'hen-peck'd Sire'[6] and by the middle of the century the hen-pecked husband was such a familiar feature of British literature and art, he gained a new nickname – the 'meacock' (a meek cock), no doubt a play on both downtrodden cockerel and the husband's emasculated state. In 1788, Robert Burns threw his hat into the ring with the particularly unlovely 'Henpecked Husband':

Curs'd be the man, the poorest wretch in life,
The crouching vassal to a tyrant wife!
Who has no will but by her high permission,
Who has not sixpence but in her possession;
Who must to he, his dear friend's secrets tell,
Who dreads a curtain lecture worse than hell.
Were such the wife had fallen to my part,
I'd break her spirit or I'd break her heart;
I'd charm her with the magic of a switch,
I'd kiss her maids, and kick the perverse bitch.[7]

'Hen-hertit' or 'hen-hearted' is a description of someone diffident and cowardly.[8] One of the earliest appearances of the phrase comes from the York Corpus Christi Plays, a medieval collection of mystery plays performed in the city from the mid-fourteenth century; in *The Tilemakers' Play*, one of the characters chastises another for his 'henne-harte', a metaphor also drawn on in 1545 by John Skelton, poet to King Henry VIII, when he wrote 'herted lyke an hen'.[9] And while the phrase was clearly known since the Middle Ages, it's difficult to establish where it came from. Cowardice is often a trait attributed to poultry in English – we have chicken-livered, poltroon (possibly from *pullus*, the Latin for young fowl), chicken-hearted, chickening out, to play chicken, chicken-shit and many other barnyard delights. In real life, however, chickens are not known for their timidity – cockerels can be truly aggressive and a mother hen fiercely protective of her chicks, but somewhere along the way the bird seems to have gained a reputation for weakness. It's perhaps telling that the early versions of the metaphor use the female noun 'hen', rather than the gender-neutral 'chicken', an indication that 'hen-hearted' may have originally been an insult that drew on women's perceived weakness and domesticity rather than any inherent quality of the farm bird.

As with many words in the English language, there are often double standards when it comes to gendered meaning. The word 'bantam', for example, was used in a leering way to describe young women but its associations weren't always negative, or at least the word acquired a plucky overtone when applied to men. In the run-up to the First World War, the British Army was desperate for fit, healthy young recruits but the minimum height requirement for a soldier was five feet three inches (160 cm). While many adults in poverty-stricken areas failed to reach these measurements due to poor nutrition, in some industrial and coal-mining areas a diminutive stature was not thought of as a sign of weakness - many of its strongest men were often stocky and tough, but short. When Alfred Bigland, MP for Birkenhead in Cheshire, heard the news about a group of local miners who had been rejected by the recruiting office, he petitioned the War Office for permission to create a special fighting unit just for undersized men. They were to be called 'Bantams' after the breed of little chickens whose males often fought well above their weight, and a word that since the 1880s had become commonly used to describe a lightweight class of boxers. News of this petite but punchy battalion soon travelled and, by 1914, around three thousand short men, previously barred from entry, had joined one of two new Bantam battalions.

Dialect words for chicken can also help us follow migrations of communities through time and space. 'Chook', a quintessentially Australian word, first appeared on Antipodean soil as 'chuckey'; in *Land, Labour, and Gold,* published in 1855, prolific English writer William Howitt, who had set sail for Australia to try his luck at gold prospecting, wrote 'they tied chuckey up in a handkerchief and rode on'.[10] By 1880, the Sydney *Bulletin* talked of 'little chookies'.[11] 'Chucky' and 'chuck', however, had long been in use across England and Scotland, most likely corrupted from chick or chicken, and used as a fond greeting.* Shakespeare included the word at least six times in his work, applying it to both men and women, including in *Love's Labour's Lost* (1598), when the princess is called 'sweet chuck'; in *Macbeth* as 'dearest chuck'; and in *Othello,* when Desdemona is asked 'What promise, chuck?'[12]

While the word seems to fall out of use in the southern counties, the north clung on to 'chuck' and 'chucky' but with slightly different regional meanings. In Scotland, 'chucky' was often used in place of the bird, as in a line from Allan Ramsay's early eighteenth-century poem 'the Priest Shot in his Fork in Chucky's Breast',[13] while across northern England, especially Yorkshire and Cleveland, 'chuck' retained both meanings – as a synonym for chicken and an affectionate greeting, as in 'Eh up,

* Samuel Johnson's *Dictionary of The English Language* (1755) records the term as 'a word of endearment'.

chuck?' or 'my chucky egg'. Yorkshire writer Emily Brontë has Heathcliff in *Wuthering Heights* (1847) asking, 'Will you come, chuck?'[14] William Howitt and many other émigrés raised in the north of England would have no doubt taken 'chuck' and 'chuckey' to their new Australian homeland during the late eighteenth and nineteenth centuries, where it eventually came to mean only the bird. Conversely, the word 'chuck' in America dropped its poultry connection but remained a term of familiarity, albeit not an especially common one.

So when did the word 'cock' acquire its boorish meaning? Cockerel is certainly an ancient word. As we saw in Chapter 1, Mohenjo Daro in Pakistan was possibly *Kukkut arma* or 'City of the Cockerel', a civilisation built at least four and a half thousand years ago. *Kukkut,* a word that may have been inspired by the cockerel's voice, eventually found its way around the world - we see *kukkuta* in Sanskrit, *kikkos* in ancient Greek, the Slavic *kokot*, Old Norse *kokkr* and even Old English *cocc*. We also know that both Greek and Roman culture conflated the cockerel with ideas about aggression and male virility. This expressed itself in many ways - from the popularity of cockfighting among soldiers to lovers' gifts. And while these ancient civilisations didn't necessarily use the word 'cockerel' as a synonym for male genitals, the artwork and ceramics from the period are often covered with motifs of men dressed as cockerels with phalluses or strange hybrids of birds with enormous, human penises.

After the Romans left Britain, we know very little about how cockerels were viewed symbolically by the Anglo-Saxons but historians have noticed that depictions of the cockerel are rare in Christian art from this time, as are other male animals traditionally linked with fertility, such as the wild boar and the stag.[15] What this means isn't clear, but the absence of these motifs may have been part of an active attempt, on the part of the early Christian church, to distance itself from old beliefs and practices that linked the cockerel with sexual potency.

By the medieval period, however, the male chicken and the penis had once again become equivalents. The slang word 'cock' – to mean penis – seems to have been in use at least as early as the 1300s, but in the longer form 'pilcock' or 'pillicock'.* The roots of 'pilcock' aren't clear but the word might have been a marriage between the Old English *cocc*, meaning the cockerel, and *pil*, an ancient word for either a rod or ball, depending on its origins. The Kildare Lyrics are a group of sixteen poems written in an Irish dialect of Middle English; dating to the mid-fourteenth century, they're ripe with satire, including one of the pieces – 'Elde' – a poem about the complaints of old age. The anonymous author bemoans:

Y ne mai no more of loue done;
Mi pilkoc pisseþ on mi schone

(I may no longer make love,
My cock pisses on my shoe)

By the early fifteenth century, the saucy synonym was used to great effect in another anonymous poem, 'I have a gentel

* We still use 'pillock' as an insult, with many of us not knowing that it means penis.

cock'. For most of the lines, the verse seems like a charming description of a young man's noble cockerel, but as the poem goes on readers are left in little doubt of its thinly disguised innuendo:

I have a gentle cock,
Croweth me day;
He doth me risen early
My matins for to say.

I have a gentle cock,
Comen he is of great;
His comb is of red coral,
His tail is of jet.

I have a gentle cock,
Comen he is of kind;
His comb is of red coral,
His tail is of inde. *

His legges be of azure,
So gentle and so small;
His spurres are of silver white
Into the wortewale. **

His eyen are of crystal,
Locked all in amber;
And every night he percheth him
In my lady's chamber.

* Inde' was indigo or black.
** The 'wortewale' is an extant word for the root or base of a cockerel's spur.

For medieval readers of the poem, the double entendre of the cockerel entering a 'lady's chamber' would have been hilarious, bolstered by other cheeky allusions such as the cockerel forcing his master to wake up in the morning. Almost every body part of the cockerel had a sexual undertone – the 'tail' was another word for penis, the coxcomb the glans. Even the spurs were believed to be a secondary sexual organ, which if removed or cauterised would render the bird infertile. Throughout the Tudor period, bawdy language and chicken-based sexual puns were *de rigueur* – Shakespeare's plays are littered with deliciously crude chicken insults and quips, such as 'Pillicock sat on Pillicock Hill'* in *King Lear* and 'I must lose my maidenhead by cocklight'** in *The Two Noble Kinsmen.*

'Cock' also had a similar, related meaning of bold, youthful energy. As one Victorian writer noted, 'The term "cock" implied *pertness*: especially the pertness of lusty and swaggering youth. To cock up the eye, or the hat, or the tail, a haycock in a field, a cock-robin in the wood, and a cock-horse in the nursery, all had the same relationship of meaning—brisk action, pert demonstrativeness.'[16] To the Elizabethan, a 'cockapert' was a saucy fellow; any red-blooded lad a 'cock' or 'cocker'. Indeed, 'cocker' and 'cock' are still used as a friendly greeting across Lancashire and parts of Yorkshire. Coquettish – a word now used only for women, meaning insincerely flirtatious – was originally a unisex concept. 'Coquet' (male) or 'coquette' (female) – from the French for 'little cock' – were in common use

* Pillicock Hill is thought to refer to the female genitals.
** Cocklight meaning both dawn and a reference to the penis.

by the seventeenth century. Double standards applied from the start – while a male 'coquet' was amusingly and enthusiastically amorous, a female 'coquette' had negative, gossipy connotations; Randle Cotgrave's 1611 edition of *A Dictionarie of the French and English Tongues* lumped together 'a prattling, or proud gossip; a fisking, or fliperous minx; a cocket, or tatling housewife; a titifill, a flibergebit.'[17]

After the establishment of the Commonwealth of England in 1649, Oliver Cromwell – a passionate Puritan – set about attempting to restore the moral and spiritual health of the nation. Along with cancelling Christmas, banning the theatre and betting sports, and prohibiting women from wearing make-up, Cromwell also targeted swearing and profanity, even warning his own armies 'Not a man swears but pays his twelve pence'. While many potty-mouthed citizens got off with a fine or a short stint in prison, some weren't so lucky. On one notable occasion, a quartermaster called Boutholmey was condemned to have his tongue bored with a red-hot iron for swearing.[18]

In the decades leading up to the Civil War, England had been in religious turmoil and had not been an easy place for nonconformist Puritans to worship freely. In the 1620s and 1630s, thousands of Puritans had fled the country to establish new colonies on the east coast of America, eager to live a life free from religious persecution. During Cromwell's campaign in the 1640s, around 10 per cent of Puritan colonists briefly returned to England to assist the Parliamentarian military effort, but the majority stayed behind and continued to practise the faith in their adopted country. With the restoration of the

Stuart monarchy in 1660, England's Puritan movement was largely eclipsed but in North America Puritanism flourished and sought to cleanse society of what it regarded as sinful practices, including gambling, drunkenness and swearing.

The English word 'cock', it seems, was too much for some American Puritan sensibilities to bear – within a hundred years the ancient noun had been replaced in polite conversation by the altogether more palatable 'rooster', a bird named after its napping place. 'Roost' comes from the Old English *hrost*, the wooden framework of a roof. Its first appearance in literature comes from 1772, and the remarkable diary of Anna Green Winslow, a twelve-year-old girl sent hundreds of miles away from her Nova Scotia family to boarding school in Boston. In a letter home, she describes dining with her uncle and a dish that 'contain'd a number of roast fowls—half a dozen, we suppose, & all roosters at this season no doubt'.

Quite when 'cock' became too uncouth for polite parlance isn't clear – Anne Bradstreet, one of the most prominent early colonial poets in North America and daughter of a wealthy English Puritan, had no compunction about using the word in her poem 'The Four Ages of Man'. 'The early cock', she wrote with not a hint of a smile in 1650, 'did summon but in vain, My wakeful thought up to my painful gain.'[19] Neither did Richard Saunders, author of *Poor Richard, An Almanac For the Year of Christ 1739*, whose annual predictions included the gloomy forecast of 'Whole Flocks, Herds and Droves of Sheep, Swine and Oxen, Cocks and Hens, Ducks and Drakes, Geese and Ganders shall go to Pot'.[20]

By the early nineteenth century, however, it seems 'cock' raised one too many eyebrows among the fine-mannered classes. Even the father of *Little Women*'s Louisa May Alcott – the educator and writer Amos Bronson – felt it necessary to change his surname from Alcox to Alcott. Amos had been born into a struggling farming family but had fierce ambitions for a better life. Tall and charming, he had managed to befriend some wealthy families who introduced him to art, architecture and high-society life. Keen to emulate their ways, Amos dropped his rural accent, adopted 'the manners of a great Peer'[21] and changed his rustic surname. Over the decades, the family had already gone through a number of different spellings – including Alcock, Alcocke and Alcox – but the pun-ridden name didn't fit with Amos's idea of a cultivated gentleman. In 1815, Amos began signing the occasional letter 'A. B. Alcott' and by 1821 the titter-inducing surname had finally been abandoned.[22]

The British, however, never dropped the word 'cock', despite its connotations, nor ever took to 'rooster'. James Flint, author of *Letters from America*, had to explain the meaning of the word 'rooster' to his English audience in 1822. On his travels around the country, he had noticed a number of new terms he'd never heard before; 'These I must call Americanisms', he concluded. Among the novel words, which included 'chores', 'raised' – as in 'brought up' – and 'bos' to mean 'master', he also included 'Rooster, or he-bird', which he defined as 'Cock, male of the hen'.[23] The campaign to expunge the word reached giddy new heights by the middle of the nineteenth century; not content with ridding the language of 'cock', refined American speakers also replaced haycocks with haystacks, weathercocks with weathervanes and cock-horses with hobbyhorses.

One of the most debated of the chicken words is 'Cockney', a term used to describe a native of London, or more precisely, someone born within the sound of Bow Bells of St Mary-le-Bow church in the City. Much has been made of the fact that 'cockeneyes' was a fourteenth-century word used to describe 'cocks' eggs', the small infertile eggs laid by a young hen, but during the same period, in 'The Reeve's Tale', one of the *Canterbury Tales*, Chaucer used the word 'cokenay' to describe a male character who chides himself for being weak:

And when this jape is tald another day,
I sal been halde a daf, a cokenay!

(And when this jape is told another day,
I shall be thought a fool, a weakling!)

By Elizabethan times, the word had come to be used as an insult directed towards city folk, people viewed as having none of the real-world experience and grit of country dwellers. Robert Whittington, grammarian in the early sixteenth century, remarked sourly, 'This cokneys [...] may abide no sorrow when they come to age. In this great citees as London, York the children be so nycely and wantonly brought up that comonly they can little good.'[24] While some have interpreted cokney or cokenay to mean 'spoiled', perhaps from the Old French *acoquiné*, a more obvious answer is perhaps staring us in the face. The English poet and writer Thomas Tusser, in his *Five Hundred Points of Good Husbandry* (1573), warned housewives and mothers that 'Som cockneis [...] are made very fooles, fit neither for prentice, for plough nor for schools.'[25] All of these career choices - apprentice, farmer or scholar - were almost exclusively male pursuits; could the word simply be a conjoining of 'cock' and 'nay' (meaning no or not), i.e. unlike a cockerel? By 1600, the word narrowed its geographical target; Samuel Rowlands' intriguingly titled *The Letting of Humors Blood in the Head-Vaine* boasted, 'I scorne (that any Youngster of our Towne) To let the Bowe-Bell Cockney put mee downe'. Less than two decades later, the travel writer Fynes Moryson observed that 'Londiners, and all within the sound of Bowbell, are in reproch called Cocknies'.

'Cockney' retained its negative connotations well into the twentieth century, but the criticism changed; Cockneys were no longer thought of as effeminate and unmanly; instead, the people and language of the East End were branded uneducated and uncouth. Speakers of received pronunciation were horrified that citizens of England's capital city were speaking with their own regional dialect and, even worse, Cockney influence was spreading. Edward Gepp, in his *An Essex Dialect Dictionary* (1923), panicked that 'Modern Cockney language has now crept in among us, and is creeping more and more, and we regret and resent it. [...] The deadening influence of London is seen for many miles out [...] the poison is in the air, and the blighting Cockney's Sahfend (Southend), Borking (Barking) and Elestead (Halestead) and the like show what we may come to. Heaven preserve us!' Thankfully, in twenty-first-century Britain, 'Cockney' is a term now brimming with pride. Few realise that the word, worn as a badge of honour, started life as a medieval chicken-inspired slur.

'Cock-a-hoop' is an idiom that also gets etymologists sparring. Many have tried to explain the origin of this curious saying, which means 'triumphantly boastful'. Two main contenders are often put forward – the first is that the 'cock' refers to a beer tap and the 'hoop' is the metal ring around the top of a beer barrel. Thomas Blount in 1670 confidently asserted that, in times past, 'the Cock being taken out, and laid on the hoop of the vessel, they used to drink up the ale as it ran out without intermission and then they were Cock-on-Hoop, that is, at the height of mirth and jollity; a saying still retained'.[26] Another contender comes from the French phrase *coq à huppe*, meaning a cockerel with a raised crest or comb, which was interpreted as displaying bravado or defiance.[27]

However, an interesting quirk of ancient pubs may provide another explanation. Along with straight roads, central heating and fermented anchovies, the Roman army also brought the idea of pubs to Britain. *Bibulium* or inns were built alongside roads and in towns to quench the thirst of troops and, eventually, locals. These taverns would display a pub sign – a wreath of leaves, wrapped around a hoop, at the end of a pole that projected out into the street, which could be seen from a great distance. The idea stuck and became the 'ale-stake' of Anglo-Saxon and medieval inns; the hoop remained a visual framing device for the pub's chosen symbol and as early as 1369 we have pub names ending in '-on-the-hoop'. A *Dictionary of Inn-Sign Names in Medieval and Renaissance England* includes 'George on the Hoope', 'The Belle on the Hoop', 'Harp on the Hoop' and a good number of 'Cok on the Hoops'.[28] 'Cock-a-hoop' may have been born out of an association with the pleasures of pub life.

Another phrase that has attracted debate is 'chicken pox', a disease that – on the face of it – has absolutely nothing to do with chickens. The word was first recorded in 1694 by talented English physician Richard Morton. In his *Exercitatio de Febribus Inflammatoriis* (*Exercise on Inflammatory Fever*), he talked about a version of smallpox 'commonly known as Chicken-Pox'.[29] While he was mistaken in thinking that the diseases were one and the same, it was clear that the name 'chicken pox' was already part of everyday speech. What isn't clear is why.

In Thomas Fuller's *Exanthemologia*, a medical reference book published in 1730, the author offered the rather cute explanation that it was after 'the smallness of the Specks, which [our Women] might fancy looked as tho' a Child had been picked with the Bills of Chickens'. In 1886, the British surgeon Charles Fagge, in *Principles and Practice of Medicine*, suggested that chicken pox might be so called because the pustules looked like chick-peas. While this provides a wonderfully visual explanation, the word 'chick-pea' only came into the English language in the eighteenth century, too late for Richard Morton's pioneering description. Other suggestions include chicken pox being so called as it makes the sufferer's skin look like plucked poultry or because of the possible sound similarity between the Old English word *giccan*, 'to itch' and 'chicken'.

More convincing, however, are two options both related to archaic uses of the word chicken. One thought is that chicken pox could mean 'child's pox' – as we've already learned, the words 'chick' and 'child' were often interchangeable, and the disease is largely one of childhood. The second suggestion is that chicken pox could be so called because it wasn't a serious disease compared to smallpox or great pox (syphilis), drawing on the sense of chicken as 'weak'.

Language, of course, evolves and some chicken-inspired words have disappeared into the ether. The rather lovely 'chickling' was in use up until the nineteenth century to mean a baby chick but only survives in the common botanical name 'chickling vetch', a pea plant viewed as inferior to other kinds of legumes. The same derogatory use of the word applied to 'chick-pea' – one nineteenth-century dictionary describes the food, a little unkindly, as 'A kind of degenerate pea'.[30] In Middle English, 'chukken' was to make a clucking noise like a chicken but by the late sixteenth century came to mean 'laughing', a sense that survives in both 'chuckle' and 'chuckling'.

A 'hennin' was a headdress worn by women of nobility in the Late Middle Ages – the classic pointed princess hat with a swooshing veil. Etymologists suspect the word 'hennin' probably grew out of the archaic word for cockerel, 'hahn', after the headwear's resemblance to a cock's comb. The painful-sounding 'cockshut time' was used to describe twilight, the hour that chickens were shut into their coop. Shakespeare uses the phrase in *Richard III*, describing 'Thomas the Earl of Surrey, and himself, / Much about cock-shut time, from troop to troop. / Went through the army, cheering up the soldiers.' 'Cocklight', conversely, meant dawn. Elizabethan audiences would also have understood the word 'ninnycock', a brilliantly evocative insult meaning a simpleton ('ninny' meant innocent or child-like).

Some eighteenth-century words thankfully left behind include 'cock alley' or 'cock lane', both unsavoury descriptions of a woman's private parts, a 'bully cock', someone who starts an argument, a 'flat cock' meaning a woman, and 'to cry cockles', which meant to be hanged.[31] 'Cockle' was a word used to describe the low, drawn-out noises made by a cockerel and thought to sound like someone being strangled.

During the early nineteenth century, two spellings of the yellow part of the egg were still perfectly acceptable – yolk and yelk. Sir Thomas Browne, writing in 1646, noted 'a Chicken is formed out of the yelk of the Egg'.[32] Yelk was particularly popular in America and proudly defended; the 'Father of American Scholarship', Noah Webster, in his 1789 *Dissertations on the English Language*, confidently claimed, 'The word *yelk* is sometimes written *yolk* and pronounced *yoke*. But *yelk* is the most correct orthography, from the Saxon *gealkwe*; and in this country, it is the general pronunciation.'[33]

The American Housewife, a nineteenth-century collection of recipes, was still insisting on using yelk not yolk half a century later – 'Eggs look very prettily cooked in this way,' the anonymous 'Experienced Lady' enthused, 'the yelk being just visible through the white.'[34] She would have also, no doubt, enjoyed an 'egg-wife's trot', a delightful alternative for a gentle walk, and assiduously avoided 'eggtaggling', a brilliant word for wasting time, especially in bad company. The last of these is thought to be a blend of egg and *taigle*, an old Scottish word for hinder or delay, and was designed to conjure up the image of a ne'r-do-well pretending to look busy as if searching for eggs.

Two of the best-known phrases are, however, surprisingly modern. The first is to describe someone manic as 'running around like a headless chicken'. It's an evocative metaphor and surprisingly absent in colourful medieval texts or Shakespearean banter. In fact, the phrase was unknown until the nineteenth century. One of the earliest references to the weird phenomenon of decapitated chickens and their reflex movements comes from an 1851 English newspaper: 'Headless chickens', noted the journalist, 'still flutter their wings and run a little.' He goes on to compare the death throes of a chicken to a piece of 'mutilated and dismembered' legislation ruined by the meddling of MPs in the House of Commons.[35] Two years later, an American report used the idiom in a way more familiar to modern audiences; the Cleveland *Plain Dealer* announced: 'The Free Soil party [...] is divided and bewildered like a headless chicken, and all through their own exertions.'[36]

The phrase seems to have been initially more popular in America than Britain, partly because of a craze around that time for gruesome sideshows featuring headless chickens. America became obsessed with a handful of incidents where farm chickens had survived decapitation. It's uncommon, though not impossible, to chop off the top of a chicken's head but leave enough of the brain stem intact for it to survive for several weeks after (a chicken's brain is behind the eye, rather than on the top of its head).

To keep the bird alive, the chicken had to be artificially fed in some way, usually via the remaining section of the neck. So curious and newsworthy, these headless chickens soon became local celebrities, with people queueing up to see the miraculous creatures and paying handsomely for the privilege. Unscrupulous entrepreneurs, keen to make a quick buck, began to try and recreate headless chickens of their own. One story from the *Alexandria Gazette*, Virginia, in 1868, highlighted the fowl deed:

> '**HEADLESS CHICKEN** - Some inhuman creature - an Italian - has been placarding the streets of the city announcing the exhibition of a living "headless rooster." The admittance charged was fifteen cents. Someone, cruel to a degree, had cut from a living fowl all of the head except the brain; had healed the wound by means of plasters, and had sustained life in the body of the mutilated fowl by introducing Indian meal into its craw by artificial means. The tortured bird lived, and must have proved a source of considerable revenue to the unnatural person by whom it was exhibited.'[37]

Famous headless fowl included the 'Martinez chicken of San Jose', which lived for three months, and the 'San Francisco headless chicken', a bird so famous it even toured New Jersey, four thousand miles from its home range. None caught the press attention more, however, than Mike, a male Wyandotte chicken that lived for a year and a half after his head had been cut off in 1945. In a botched attempt to kill Mike, Colorado farmer Lloyd Olsen had removed the bulk of the chicken's head with an axe but crucially missed the jugular vein and left most of his brain stem intact. Olsen's cack-handed butchery skills left Mike still able to perch and walk, and even try to crow (although he produced only a faint, throaty gurgle), and so Olsen decided to keep Mike alive by feeding him through the throat with an eyedropper. News of the miraculous Mike spread and he toured the country, appearing in numerous sideshows, newspapers and photoshoots. At the height of his fame, Mike was reputed to be earning Farmer Olsen over $4,000 a month, not bad for a bird with no head for figures.

Another surprisingly modern phrase is 'pecking order'; it's used to explain human relationships, or group dynamics, and means a hierarchical chain of command. Chickens are flock birds but not all are born equal. Flocks have strict hierarchies or 'pecking orders', a term first coined in the early twentieth century by the Norwegian zoologist and comparative psychologist Thorleif Schjelderup-Ebbe. Thorleif had been obsessed with chickens since childhood and, from the age of ten, began making detailed notes about his own flock's behaviour. In particular, he was interested in the relationships between individual birds and how chickens 'knew their place'. He observed that a bird's ranking in the group emerged from fights over food and that each chicken knew who was above and below it in the system of ranking. Dominance was reinforced by a sharp peck - often to the head - hence the term 'pecking order', and both males and females had separate dominance hierarchies.

More recently, researchers have studied just what makes a chicken more likely to rise through the ranks in terms of pecking order. Factors such as age, body size and shape, overall health and social experience seemed to be important factors in establishing a chicken's status within a group,[38] but inheritance also plays a role. In flocks that have been established for years, pecking order rank is passed down through the generations. The first daughter of a dominant hen, will 'inherit' her title. The daughter of the second in command will inherit this role, and so on. If the dominant hen has more than one surviving daughter, however, her offspring will pull rank over the second in command.

The purpose of the pecking order is to maintain group cohesion and regulate access to resources. The pecking order affects lots of different aspects of a bird's life, including which order they eat and drink in, where they'll roost and choice of mate. Stronger, more dominant chickens will get their pick of tasty treats, water, dust baths and partners, but with great power comes great responsibility. The head cockerel has to play the role of flock protector, keeping a lookout for predators and issuing warning calls if danger looms. If other chickens brawl, the head cockerel will try to intervene. In the absence of a cockerel, an alpha female will step into the role. Like in a medieval court, the head cockerel will surround himself with the more dominant hens, while the lower-ranking males and females must make do with life on the periphery. And, as with all dynasties, a high-ranking individual can be usurped by a new upstart or be forced to relinquish his or her position due to age or ill health.

Once a pecking order has been established, communication between a group of chickens is remarkably sophisticated. Chickens are chatty creatures, but we've only just really started to understand their fowl language. Between the 1950s and 1980s, Nicholas and Elsie Collias at the University of California identified over twenty different chicken vocalisations and their likely meanings. The sounds produced by chickens were found to be very different depending on the situation, from a quiet but high-pitched 'eeeeee' when faced with an aerial threat such as a bird of prey to the enticing 'dock dock' a cockerel makes when revealing a titbit of food to a hen.

Only with the development of sophisticated digital recording techniques in the 1990s, however, have scientists been able to establish a more accurate connection between distinct events and chicken 'language'. Researchers at the Macquarie University in Sydney, Australia, created virtual environments for chickens using television screens. The birds were played different scenarios to test their responses when they saw predators on the screen, such as a raptor or racoon, but also videos of 'friends' and flock competitors. The chickens' vocalisations were recorded, and the results were rather surprising. Not only were chickens making sounds that conveyed specific information but their flock mates could understand this information without needing to see the particular stimuli. To do this, chickens' brains have to be able to create a mental picture and respond accordingly, contradicting previous assumptions that chickens could only communicate simple vocalisations such as being scared or aggressive. There'll be more about chicken intelligence in the final chapter.

What's even more fascinating, however, is just how Machiavellian chickens can be; studies have shown males will make false food calls to attract females nearby, and that hens will begin to ignore a cockerel who 'cries wolf' once too often. Males will also cluck more quietly while trying to woo a female if a competitor is near, so as not to alert any potential rivals to her affection. Perhaps most surprising is the cockerel's ability to change his alarm cries depending on who's in the vicinity; research has shown that if an aerial predator flies above a male chicken, he's more likely to sound a warning if he's near a place of refuge or near another male rival. Alarm calling is a survival conundrum - although the cockerel will increase the likelihood of his flock being protected, by drawing attention to himself he's also putting his own life at risk. To solve this problem, cockerels have adopted at least two strategies - one is to make sure they're near a hiding place before they raise the alarm and the other, more cunningly, is to ensure a male rival is close by in the hope he'll get eaten instead.[39] Turns out the metaphor 'hen-brained' has never been further from the truth.

5

PETS

Hen Fever

Cochin

At the end of the seventeenth century, John Smith wrote *England's Improvement Reviv'd* - an ambitious book on husbandry and horticulture - aimed at the landed gentry. In it, Smith suggested that a landowner might enjoy redesigning their agricultural land 'as well for Pleasure as Profit'. He painted a glorious picture of a pastoral idyll, an estate that might include 'several Orchards and Gardens, with Fruit, Flowers and Herbs both for Food and Physick, variety of Fowl, Bees, Silk-worms, Bucks, Does, Hares, and other Creatures [...] Also Fish-ponds and Streams of water stored with many kinds of Fish, and stocked with Decoy-Ducks; And the Use and Vertues of all the Plants in this Garden of Pleasure.'[1] Once these elements were in place, the owners of these rural arcadias could trip through their 'little Theatres of Nature', masters of all they surveyed; all the flowers, trees, crops and domestic animals in this arrangement were part of an earthly Eden, for them to use and enjoy at their leisure.

The fashion for these 'ornamental farms' or *fermes ornées* gained popularity both in Britain and France through the eighteenth and early nineteenth centuries. This was the era of Romanticism, a movement that celebrated the deep beauty of the natural world, folk culture and the simplicity of rural life; the romanticised farm, complete with its wholesome animals and peasants, embodied the virtues that many felt were being eroded by industrialisation and the application of scientific reason to every facet of life. Marie-Antoinette famously created her *Hameau de la Reine* or Queen's Hamlet in the 1780s in the park at Versailles, a fake semi-derelict village complete with farmhouse, mill, dairy, aviary and barn. The *hameau* even had its own little farmer, Bussard, who patiently tended to the queen's livestock and crops while she and her friends meandered through, taking in the sights and sounds of the faux-countryside.

A *ferme ornée* wasn't complete, however, without a hen house. Many of the great houses built or adapted during this time persuaded some of the biggest names in architecture to design first-class accommodation for their fowl. In 1794, the same year he created the magnificent rotunda on the Bank of England in London, Sir John Soane also drew up plans for a hen house for the Earl of Hardwicke's Wimpole Hall in Cambridgeshire, a miniature classical stone facade complete with nine perfectly arched nesting boxes. When the 6th Duke of Bedford inherited Woburn Abbey in 1802, he tasked the celebrated designer Humphry Repton with creating a coop to house his ever-growing menagerie including 'pigeons, chickens, and heaven knows what birds', according to one German visitor.[2]

Lord Penrhyn, at Winnington in Cheshire, no doubt built the grandest hen house – a 40-metre pavilioned 'poultry palace' – but perhaps none was so eccentric as the second George Durant's chicken pyramid. The Durants of Tong, Shropshire, epitomised the unbridled spending and indolence of the late Georgian aristocracy. After a scandalous affair with a senior politician's wife, George Durant worked as an army paymaster on an expedition to Havana in 1762, amassing enough of a fortune from looting and slavery to buy a large estate and rebuild the mansion and grounds in the latest style. His equally lascivious and very odd son, George Durant II, carried on his father's legacy in both spirit and practice, constructing numerous bizarre buildings to house the estate's livestock including a castle for cows, an elegant piggery and a 20-foot pyramid coop.

The towering neo-Egyptian aviary was carefully built from hand-made bricks and coursed sandstone, with small diamond-shaped entrance holes for the chickens. To add to the madness, some of the bricks were inscribed with strange mottos for the birds, such as 'Live and let live', 'Scrat before you peck' and the inexplicable 'Teach your granny' and 'Can you smell'.[3] Durant took the idea of *fermes ornées* one step further, however, viewing not only his livestock but his estate workers as part of his rural dominion and living a life of *droit de seigneur* debauchery. Fathering close to forty offspring, including twenty legitimate ones, Durant was reputed to have sired a child in every cottage on the Tong estate. As befitting a man who viewed most things in his purview as both playthings and property, he gave his illegitimate children names more suited to pet chickens than progeny, including 'Napoleon Wedge', 'Columbine Cherrington' and 'Cinderella Greatback'.[4]

Even Queen Victoria wasn't immune to the charms of keeping chickens as pets. Two days before Christmas 1843, *The*

Illustrated London News printed a gushing double-page special celebrating 'The Queen's Poultry' at Windsor. Its aim was to 'show the readers a fair and seasonable picture of our gracious and nature-loving Queen, and of her winged favourites in the royal poultry yards of Windsor'.[5] Queen Victoria's predecessor, King George III, had built a private farm in the large park to the east of Windsor Castle, including poultry yards to provide meat and eggs for the family. Soon after her accession, Victoria commissioned a vast new fowl-house, designed and built by Windsor's go-to royal builders, Messrs Bedborough and Jenner.

The 'semi-gothic building of simple and appropriate beauty' included a large central pavilion with wings and roosting houses for the birds to lay, nest and snooze. No expense was spared; 'the chambers are spacious', enthused *The Illustrated London News*, 'airy, and of equal and rather warm temperature [...] their nests are made as far as possible to resemble the dark bramble covered recesses of their original jungles.' Crucially, the chickens enjoying this pampered accommodation were not your ordinary mixed breed or 'dunghill' chickens, but 'rare and curious' specimens. The queen was clearly enchanted by her unusual flock, spending hours down at the fowl-house 'retiring from the fatigues of state'. The hen home wasn't just a practical space but a place where the queen could find 'great relief in the simple pursuits of country life' and recover 'those higher powers which find their best, if not their only home, in nature'.

Throughout history, keeping chickens as pets was largely the preserve of the wealthy. Farm animals were traditionally expected to earn their keep, but it seems little was demanded from Victoria's pet poultry. Her outlandish brood included: 'several frizzle fowls, remarkable for their white, silky, hair-like feather and their black skins'; dozens of problematic Java bantams, the feisty cockerels of which constantly smashed and ate the females' eggs; and some 'curious "crosses" with grouse birds'. Chickens and other game birds had, occasionally, been known to mate if their paths crossed in the wild but rarely spawned any fertile offspring. Victorian scientists and chicken breeders, however, were keen to find out how far these hybrid partnerships could go, matchmaking domestic chickens with almost every close cousin they could capture.

Successful crosses included chickens and pheasants, chickens and quail, chickens and grouse and, remarkably, even chickens and peacocks. In almost all cases, the resulting chicks lost much of their domestic hen-like docility. Charles Darwin's interest was especially piqued by the results of one notable breeder; in his *The Variation of Animals and Plants Under Domestication*, Darwin noted: 'Mr. Hewitt, who has had great experience in crossing tame cock-pheasants with fowls belonging to five breeds, gives as the character of all "extraordinary wildness".'[6] While most couplings produced jittery, sterile birds, the occasional match resulted in fertile offspring, especially between hens and junglefowl. Indeed, it was many of these early trials that later gave Darwin the confidence to declare the red junglefowl as the definitive ancestor of the modern chicken, based on the fact that the poultry pairing was one of the only to reliably produce fertile young.

However unusual the queen's 'chick-grouse' were, the real stars of the coop were her newly acquired 'Cochin-China fowls', birds of such a gigantic size and upright stature they had been nicknamed 'ostrich-fowl' by the press. The birds had been gifted to the queen in the early 1840s by the controversial naval officer and explorer Sir Edward Belcher,* who had picked up the poultry on his way back from a protracted six-year voyage around the world. Although the breed was initially known as 'Cochin-China' – the colonial name for Vietnam – it's not clear where exactly in Asia the birds had been brought on board. The queen's birds looked not unlike Malay chickens, a large, heavy-boned chicken found widely throughout north India, Indonesia and Malaysia, where it was kept for its considerable fighting prowess and towering stature.

The queen was enthralled with her new Cochin-China chickens and sent fertile eggs to royal households across Europe in the hope they too could share her new passion for pet poultry. Initial hatching attempts didn't go too well, however, according to reports. Even with the benefit of their sumptuous royal surroundings, Queen Victoria's Cochin-China cockerel quickly

* Belcher was one of the most divisive characters in the Royal Navy. While he finished his career as an admiral and was praised for his scientific endeavours and intrepid voyaging, he was by all accounts also a tyrant and a terrible husband. Crews regularly complained of Belcher's brutal treatment on board, while Belcher's wife filed for legal separation after he infected her with venereal disease, twice.

expired and the hens only managed to lay a handful of eggs. A Dorking cockerel was procured as a hasty replacement, a tough, beefy bird thought to have been brought across to England by the Romans, and the resulting crosses looked convincingly similar to the original Cochin-China birds, their hybrid heritage only given away by the peculiar Dorking trait of possessing an extra toe.[7]

Just a few years after Queen Victoria had welcomed her Cochin-China birds, another breeder – farmer and entrepreneur Alfred Sturgeon – managed to get hold of a small flock of enormous fluffy Chinese chickens fresh from a tea clipper moored in London's dockside. In a later letter, he described the moment his assistant first clapped eyes on the remarkable fowl: 'I got them in 1847, from a ship in the West India Docks. A clerk we employed at that time happened to go on board, and, struck by the appearance of the birds, bought them on his own responsibility, and at what I, when I came to hear of it, denounced as a most extravagant price—some 6s. or 8s. Each!' These birds looked different from the queen's Cochins – they were equally large but rounder, softer and downier, with wonderfully feathered legs – and were quickly named 'Shanghai' fowl after their homeland. The name didn't catch on, however, and these novel birds also became, confusingly, known as Cochins, despite having little in common with the birds roosting in Her Majesty's coop.

Sturgeon took his small clutch of 'Shanghai Cochins' back to the family farm in Essex but almost as soon as the birds arrived, disaster struck. 'Judge of my terror,' he later wrote, 'after my extravagance, when I found a younger brother had, immediately on their arrival, killed two out of the five, leaving me a cockerel and two pullets; nor was my annoyance diminished on hearing him quietly remark that they were very young, fat, and heavy, and would never have got any better!' Nor was this the end of Sturgeon's misfortunes. The surviving cockerel died shortly afterwards, and a calamitous incident involving some unruly puppies and young Cochin chicks left Sturgeon little choice but to try desperately to procure some additional birds.

Alfred Sturgeon persisted, however, as did a handful of other breeders including Mr Punchard, a friend who had taken a liking to the breed and bought a couple of birds from Sturgeon for himself; and a Mr Moody of Hampshire, who had acquired twelve hens and two cockerels from another ship arriving from China. By 1850, there were enough new birds to present the 'Shanghai Cochins' at Bingley Hall in Birmingham, a spectacular one-and-a-quarter-acre venue built only a few months earlier. Bingley was the first purpose-built exhibition hall in the country, a cathedral to Victorian optimism and prize-winning animals.

According to *Wright's Book of Poultry*, penned just a few years later, this was the moment the public perception of chickens shifted from backyard scavengers to valuable pets: 'Previously, very few people indeed except farmers kept fowls, and those only scrubs or mongrels; and one or two shows which made attempts to attract public interest, were only looked upon with a good-natured contempt. In 1850 Cochins were exhibited at Birmingham and changed everything. Every visitor went home to tell of the new and wonderful fowls, which were as big as ostriches, and roared like lions, while gentle as lambs; which could be kept anywhere, even in a garret, and took to petting like tame cats. Others crowded in to see them, and the excitement grew, and even the street outside the show was crammed.'[8]

Fowl mania or 'hen fever' hit both Britain and America hard. Or at least those with enough money to indulge their desire for prize poultry. Punchard sold out of his Cochins at the Birmingham exhibition at the exorbitant price of £5 for three birds (about a month's wages for a skilled tradesman), but Sturgeon was even cannier. He held on to his birds for another year, despite numerous offers to buy them, and then sold 120 birds – in a heated auction – for £609, over three times Punchard's price.[9]

Almost everything about the Cochin chicken appealed to Victorian aspirational sensibilities. Not only had pet poultry been given the royal seal of approval by Queen Victoria but these mighty creatures were bigger and more glamorous than their scrawny farmyard cousins. This was the age of the agricultural improver, the landed gentleman farmer who relentlessly pursued scientific principles of livestock breeding to produce meatier, woolier and milkier beasts to feed the burgeoning urban population. If it could be done with the larger farmyard animals, why not the humble chicken? Could these 'China monsters', wondered a *Times* journalist in 1853, hold the key to the poorly performing 'present state of our poultry market'?[10] 'If rich amateurs had not lavished their money upon the turf,' he mused, 'we should never have had such good horses commonly available: and the same may be said of shorthorns and southdowns – of prize sheep and priceless pigs.'

Victorian Britain was also fascinated with 'exotica', trophies from those parts of the world rapidly being swallowed up by imperial trade and colonisation. Asia – with its rich resources, spices, silks and other valuable goods – had long been a target for European political and commercial interests. In 1599 the East India Company had been formed to forge its own trading deals with Asia but had initially struggled to compete with established Portuguese and Dutch interests. By the middle of the nineteenth century, however, Queen Victoria's empire was riding high. After France's capitulation in 1815, at the end of the Napoleonic Wars, Britain emerged as the crowning victor. Without any serious European nation to rival her power, Britain set about expanding her imperial holdings, ably assisted by a formidable navy. It soon controlled most of the key maritime trade routes including those destined for China, Burma, Malaysia, India and other far-flung destinations. Few souvenirs were as admired or as transportable as a pair of exotic live chickens; even better if they could earn their keep along the way, providing eggs and entertainment for the crew.

Some people were so desperate to get their hands on the new fluffy Cochins, they tried to acquire their own directly from Shanghai, without a solid grasp of what to buy or even how to transport the poor birds back home. As one English journalist of the time noted, with not a little bemusement, 'many persons have made repeated efforts to procure fresh importations from China, sometimes by their own endeavours [...] They are, however, said to be difficult to meet with even in the far East, and hundreds are brought over which have little or no resemblance to the kind so much admired in England. Procuring the right breed at Shanghai, is not the only difficulty to contend with [...] The trial to the fowls of tossing about on the sea for months together, shut up in coops and often washed over by the waves, is one under which they die off by the half-dozen in a single night. Quite lately, a small number selected with great care were nearing the Cape in safety, and with an excellent chance of arriving here in good health, when some little animal on board made its escape and sucked the blood of all the "lovely Cochin-Chinas" in one night. Another lot, within a fortnight of the end of their long journey, disappeared mysteriously, sacrificed perhaps to some unusually delicious tureen of soup, and were never heard of more. Such are a few of the casualties which beset the briny path of the poor emigrants.'[11]

Punch, the weekly satirical magazine, took great pleasure in lampooning the new craze for prize poultry, a trend they blamed on pushy privileged ladies demanding the latest outlandish pet. Numerous column inches and cartoons ridiculed the steep prices paid for exotic chickens and the idea of cossetting an ordinary farmyard animal. One cartoon showed a wealthy wife taking her beloved Cochins for 'A Constitutional Walk', with the birds on dog leads, while shouting at her long-suffering husband, 'Dear Dear, it's coming on to rain. Run James! Quick, and fetch an umbrella and two parasols. I'm afraid my poor dear Cochins will get the rheumatism!'[12] In another edition, a fictional anecdote told the story of another browbeaten husband: 'A Lady living at

Peckham Rise has nearly ruined her husband by the enormous prices she has been giving for Cochin-China fowls. The poor fellow is always pointed to in the neighbourhood, so the story goes, as the "Cochin-China-pecked-husband."'[13] Readers were, no doubt, rolling in the aisles at high society's foolish fancy.

The staggering amounts of money changing hands for Cochins didn't go unnoticed. In fact, they seemed to inflate the prices of the few other 'pure breeds' that die-hard fans had previously maintained in relative obscurity. As one American breeder recalled of the 1852 Birmingham poultry show, 'a single pair of "Seabright Bantams," very small and finely plumed, sold for $125; a fine "Cochin-China" cock and two hens, for $75; and a brace of "White Dorkings," at $40. An English breeder went to London, from over a hundred miles distant, for the sole purpose of procuring a setting of Black Spanish eggs, and paid one dollar for each egg. Another farmer there sent a long distance for the best Cochin-China eggs, and paid one dollar and fifty cents *each* for them, at this time!'[14]

Chickens had suddenly become valuable as pampered ornaments, rather than fighters or providers of protein. In response, a number of poultry enthusiasts began to turn their attention to creating or perfecting other interesting chickens. Getting any agreement on what constituted a new breed, however, wasn't straightforward. Since the chicken's dispersal from Southeast Asia at least four thousand years ago, hundreds of new types had emerged over the centuries; some chickens had evolved as a result of adapting to different climates or environments, others had been deliberately bred for specific behaviours or aesthetic quirks. Many had simply accidentally interbred in the freedom of feral life, producing a kaleidoscope of traits.

From its original junglefowl body shape, the chicken had morphed to include breeds that had elongated or dumpy legs, frizzled feathers or bald patches, crested heads, combs in a dozen different shapes, silky down or hard feathers, and long flowing tails or none at all. Within these breeds, numerous more varieties emerged – feather colours ranged from the darkest blacks through golds, reds and greys to snowy-white, in a whirlwind of patterns from striped to mottled, spotted to rippled. Even body sizes extended from the monstrous to the minuscule and everything in between. With judicious matches, nineteenth-century breeders soon found they could further enhance certain traits and remove others, create new palettes of colours, shrink or blow up a breed, and even change millennia-old temperaments from aggressive to entirely biddable.

After cockfighting was banned in England in 1849, the competition between breeders of game chickens also needed to find a new outlet; the battleground simply moved from the sawdust pit to the exhibition ring. Writing in the middle of the nineteenth century, William Tegetmeier, the poultry expert and friend of Charles Darwin, acknowledged the centuries of effort that had gone in to perfecting the cockfighting fowl: 'There cannot be a doubt', he wrote grandly, 'that the superiority of the Game Fowls bred in England has been entirely due to the practice of cockfighting which was extremely indulged in by all classes of society until the comparatively recent legal enactments [...] Those cocks have proved the strongest, most active and courageous, and have stricken down their antagonists in the pit [...] and the ultimate result has been that the English Game fowl is unequalled in the elegance of form, and is universally regarded as the highest possible type of gallinaceous beauty.'[15]

The ban on cockfighting initially panicked poultry keepers, who worried that the bird might go soft without the occasional 'trial of quality' that the cockfighting ring was thought to provide. As one proponent of the sport worried, 'Were the race horse not permitted to run, how would breeders discover and root out any blemish of character or form? The Game Fowl among poultry is analogous to the Arabian among horse, the high-bred Shorthorn among cattle, and the Greyhound among the canine race.'[16] Faced with the prospect that their truculent birds might lose form, a number of keepers formed a society to keep the breed at peak fitness. Competitors could now enter their prized game chickens in poultry shows and win categories such as 'Likeliest Fighting-Cock' based on a strange combination of how attractive the bird looked and how likely it was to give a fictitious opponent a pasting.[17]

The variety and geographical distribution of chickens was dizzying and a thorny challenge for anyone wanting to try to organise the birds into some sort of taxonomical order. At the end of the eighteenth century, the French naturalist Georges-Louis Leclerc bravely attempted to list the assortment he had encountered in person, or heard of through reports from colleagues and long-distance travellers. It soon became apparent that any kind of meaningful analysis of type or origin proved impossible. Lots of countries, for example, had similar versions of miniature chickens, often no bigger than a pigeon; the 'Dwarf hen of Java', for instance, bore a strong resemblance in many of its features to the 'Little English Hen', the 'Dwarf Hen of France' and the 'Small Hen of Pegu'.*[18]

Equally, Leclerc couldn't work out whether the 'Bantam Cock', a small but feisty Indonesian bird with feathery 'boots', was any relation of the 'Rough-footed Cock of France' or the 'English Dwarf Cock'. The titles given to certain types of chicken also added to the confusion, with many breeds being named after countries or cities that bore no relation to their actual heritage. The 'Hamburg', for instance, a slim but prodigious egg-laying chicken, had taken its name from the German city, but in reality had a muddled past and had also been claimed by Holland, Belgium and Russia as one of their own.[19] To add to the confusion, Yorkshire and Lancashire farmers had also been busily breeding Hamburgs as far back as the early 1700s, but called them Pheasants or Mooneys instead.

This lack of consensus had never really bothered poultry keepers. At least, not until the craze for exhibiting and buying fancy chickens took hold. Suddenly, there needed to be some consensus among competitors and judges as to what constituted a particular breed. A prize was only worth winning if everyone

* *The Burmese city of Bago.*

competing could agree on the rules. In 1865, Tegetmeier once again put his stamp on the poultry world and produced 'The Standard of Excellence in Exhibition Poultry', a brief guide to a handful of breeds upon which he and his esteemed friends could broadly agree. It described only nine breeds, each of which was only permitted to have a handful of colours with enticing paint-chart names such as Buff, Lemon or Silver Cinnamon.

As well as the fashionable Cochin chicken, Tegetmeier also set out the standards for Malays, Game, Dorkings, Hamburgs, Spanish, Polish, Bantams and an even newer breed, fresh from America, the Brahma. The Brahma, despite its name, had been developed in the United States in the 1840s by experienced breeder George Burnham, from a cross between a large, fluffy Cochin and a tall Malay game fowl imported from Bangladesh. At first they were known as 'Grey Shanghae' after their subtle silver colouring. In a deft move to promote his stock, Burnham sent Queen Victoria nine of his best birds to boost the royal hen house. Victoria was cock-a-hoop with her poultry presents and not only sent Burnham the warmest of thank-you notes but also shipped a comely portrait of herself in return. The press went wild. Newspapers lauded Burnham as the 'most successful poultry raiser in America', adding that his 'beautiful birds are creditable to him as a breeder' and 'fit for a queen'.[20]

Almost immediately, orders began flooding in from across America, Canada and Britain from collectors desperate to get their hands on these birds of 'extraordinary size and fine plumage'. Sensing there was some money to be made, Burnham redoubled his breeding efforts and by the summer of 1853 had enough birds to ship forty-two prize Brahma to England, for which he was paid the handsome sum of $870 (the equivalent of around $30,000 in today's money). Three of Burnham's Brahmas had been promised to John Baily, a fowl enthusiast from London, who later that autumn exhibited them at the poultry show in Birmingham. Much to Baily's surprise and delight, the Brahmas were instantly resold for the equivalent of $250 each, a tenfold increase in value in just three months.[21] Others soon began to replicate Baily's success, buying fancy chickens and eggs and then quickly selling them on in a rising market.

By 1855, Burnham's canny Brahma business had raked in $70,000, around $2.2 million at today's value. But the bubble was about to burst. Like so many other speculative schemes – tulip mania, the South Sea Bubble, railway fever – it climaxed and then crashed, spectacularly, leaving hundreds of investors in financial meltdown. As Burnham himself confessed: 'The nobility tired of the excitement and the people of England and the United States began to ascertain that there was nothing in this "hum".' Fancy breeds had flooded the market, people grew tired of the novelty, and the 'star of Shanghae-ism began to wane'; 'Never in the history of modern "bubbles",' he continued, 'did any mania exceed in ridiculousness or ludicrousness, or in the number of its victims surpass this inexplicable humbug.'[22]

Some men were ruined. As one desperate speculator wrote in a letter to Burnham: 'I have got on hand three hundred of the Shanghae devils! What can I do with them? There is a neighbor of mine (a police-officer), who has got stuck with a lot of "Cochin" chickens, which he swears he won't support this winter; and he has at last advertised them as stolen property, in the faint hope, I suppose, that some "green 'un" will come forward and claim them. You can't get rid of these birds! It is useless to try to sell them; you can't give them away; nobody will take them. You can't starve them, for they are fierce and dangerous when aggravated, and will kick down the strongest store-closet door; and you can't kill them, for they are tough as rhinoceroses, and tenacious of life as cats. Ah! Burnham, I have never forgiven the man who made me a present of my first lot! Do you want what I've got left? Will you take them? How much shall I pay you to receive them? Help me out, if you can.'[23]

Through modern eyes, the folly of 'fowl mania' seems absurd and yet the craze epitomised the Victorian attitude to animals. The fashion for keeping pets had become widespread at the beginning of the nineteenth century. The Victorians were obsessed with consumerism and what it could achieve. Buying things, including pets, was as much about expressing social status as it was about shopping for necessities. Creatures were goods that could be acquired, traded and displayed in the same way people today show their wealth and ambitions through their cars, clothes and houses. Pets – those animals kept for company or entertainment, rather than food or labour – would often reflect the Victorian preoccupation with status; not only were expensive, rare and exotic pets the most highly regarded, they were a means by which their owners could reinforce their place in society. The sheer cost of buying, owning and looking after certain pets also said something about a family's disposable income.

Many working-class people owned mongrel dogs, wild birds and other creatures, however, and so a defining feature of Victorian pet-keeping became the importance of breeding. Pedigrees and lineage became a way of separating out the 'special', important pets, the logic being that only those owners with the education, critical judgement and money could afford such rarefied beasts. For the aspirational middle class and the comfortable upper class, the pure-bred animal - including the fancy fowl - reflected the high status of the owner, while mongrels and 'dunghill' chickens belonged in the gutter.

The Victorian period also witnessed an upheaval in scientific thinking and people's relationship with the natural world. The work of Charles Darwin, and others, opened up questions about human evolution and our direct kinship with animals; while some people welcomed the conversation, many others found it unsettling. It's perhaps no surprise that this was the era that saw both the rise of the animal welfare movement and an increase in society's obsession with capturing, training and manipulating creatures to bend to human will. Zoos, menageries, museums and circuses revelled in man's domination over the natural world; the curious Victorian obsession with taxidermy took this idea to ridiculous lengths: making creatures pose in unnatural or amusing positions kept them disciplined and under human control for eternity.

The fowl mania crash of 1855 burned many an opportunistic investor, but a handful of onlookers had seen it coming. Two years earlier, the poultry correspondent for *The Field* magazine had predicted 'as the tulip and the dahlia, they will have their day and sink into the category of by-gone whims and fancies'. He had also made the point that, while all this fancy poultry breeding was producing some fascinating and beautiful varieties, perhaps people's focus should be on a potentially greater prize. 'The most important property of the fowl is to supply food for man', he argued; 'that species is best which supplies the most eggs of the best quality, and the best flesh for the table.' Scrolling through the new and dazzling varieties on offer, he suggested that none ticked all the boxes of the perfect commercial chicken. The Cochin was big and laid lots of eggs, but its meat wasn't as good as the Dorking. The Spanish was a good layer but a terrible 'sitter' and an even worse mother. Game fowl laid lovely eggs but were so bad-tempered no one could get near them in the poultry yard. And the Polish crested chicken, though deeply glamorous with its crested pom-pom head, laid inferior eggs, 'is a bad sitter, and a small bird'.[24] Why doesn't anyone, he wrote with exasperation, experiment with crossing some of the new varieties 'so as to obtain, in our fowl, most of the best qualities of delicacy of flesh, quantity of eggs, aptitude for incubation, and good temper'?

Other commentators were equally irritated by some of the more decorative but ultimately useless chicken traits admired at poultry shows. 'A fowl is not materially the more precious', fumed a writer in *The Times*, 'for being "gold" or "silver pencilled", "white crested" or "double combed", though "double breasted" if procurable might be an eligible quality to introduce. One variety, we see, styled "dumpies" or "bakies" attracted great attention for the shortness of their legs; but we scarcely understand the advantage of this feature, unless indeed, they will go into a smaller saucepan.'[25]

Fierce critics of the fancy poultry world, however, had failed to see what was happening in front of their eyes. For the first time in history, breeders and owners of poultry had started to appreciate the vast range of chickens that existed across the world, not just in their own backyards. Through their constant matching of different pairs of fowl, poultry keepers had also begun to *really* understand the science of selective breeding. Over the space of a few years, through the process of picking out certain characteristics, choosing the parents who best demonstrated those characteristics, and then picking the best offspring from these parents, chicken fanciers had produced dozens of new breeds and new varieties, each with its own unique set of traits.

By the beginning of the twentieth century, there were hundreds of new fancy chicken breeds in existence. While the fancy poultry world remained passionate about the fluffiest, most dazzling, gigantic or perfectly petite specimens, a number of poultry keepers were starting to think about what they wanted each breed to *do*. For anyone interested in breeding birds for the table or eggs, it was time to abandon the feather and frills approach and instead concentrate on an entirely new set of standards. Only this time it was a competition no chicken would ever want to win.

6

LAYERS

Battery Life

White Leghorn

The hen's egg is one of nature's most ingenious creations. Built from the inside out, each layer of the egg is added as it passes through the chicken's body. A fertilised egg, once laid, is also the perfect self-contained home, supplying all the nutrients and protection a growing chick could ask for. Every hen is born with thousands of immature yolks or ova. When she is ready to lay, her body releases a tiny yolk into the oviduct, a long spiralling tube in the hen's reproductive system. As the yolk grows and moves its way slowly down the oviduct, the egg white or albumen starts to form around it, followed by the shell membrane. The egg is still soft at this stage and passes into the uterus or 'shell gland'; here the egg will stay for about twenty hours, getting its finishing coat of hard calcium carbonate. For almost the entire journey, the egg travels pointy end first, and gently rotates like a bullet spinning in a barrel of a gun, but in the last hour the egg is flipped over and laid wide end first.

Which begs the question, why are hens' eggs shaped the way they are? Birds' eggs come in a huge variety of shapes and sizes, from the near-perfect sphere of a hawk-owl to the long, thin torpedo egg of the little grebe. Some eggs are also markedly conical, like the guillemot's. Hens' eggs are somewhere in between – neither elliptical nor perfectly spherical, they're satisfyingly round but slightly more pointed at one end. Over the years, scientists have tried to come up with a way of explaining this variety. One interesting theory is that egg shape is influenced by where a bird species lays its eggs. Birds who lay eggs on flat or sloping surfaces, such as clifftop species, tend to have more conical eggs, while birds who lay in cup-shaped nests have more spherical eggs. Conical eggs are less likely to roll away if they are accidentally nudged, rolling in a circle rather than trundling off.[1]

Another theory is that different egg shapes can be explained by how many eggs a mother lays and needs to keep warm during incubation. The larger the clutch size, the more necessary it is for the eggs to fit together in a small space without large air gaps. Conical shapes fit together more efficiently in a circle formation than spheres, especially if the small ends point inwards, like slices of a cake. Other researchers point to the fact that perfectly spherical eggs have the lowest possible ratio of surface area to volume and therefore need less calcium to form, suggesting round eggs might be an adaptation by birds that have diets naturally low in this essential mineral.

More recently, however, a fascinating and ambitious study of thousands of egg shapes from across the avian world revealed a surprising new possibility. The evolutionary biologist and behavioural ecologist Mary Caswell Stoddard, assisted by her team at Princeton University, took on the mammoth task of analysing almost fifty thousand eggs from approximately fourteen hundred different species of birds. Eggs were graded according to how spherical, elliptical or conical they were, and the team found that avian egg shape seems to be related to how well a bird can fly. Birds who are strong, frequent flyers, and travel long distances, often have longer, narrower and more pointed wings. Birds who are less frequent fliers or tend to stick to home territories have shorter, broader and more rounded wings. The study found that birds with narrow, pointy wings lay narrower, pointier eggs. Birds with broader, more rounded wings - such as chickens - lay more rounded eggs. It's possible that in the process of evolving their muscular, streamlined bodies, 'frequent fliers' also evolved narrower oviducts, capable of only passing longer, more streamlined eggs.[2]

The hen is clearly no high-flyer, so why are chickens' eggs pear-shaped and not perfectly oval? It's an interesting, and little understood, phenomenon. Pear-shaped eggs are actually easier for a hen to squeeze along its vent. Most of the time hens lay their eggs wide end first, a quirk even noted by the ancient Greek philosopher and polymath Aristotle. It might sound counter-intuitive to push anything out of an opening wide end first but, like squeezing an apple pip with your fingers, applying muscular pressure to the pointed end of an egg squirts it forward more easily. It used to be thought that the wide end of an egg was also its toughest, meaning there was less chance of breakage when it landed, but current research doesn't back this up. In fact, of the two ends of an egg, the pointier end is slightly better at resisting compression. Both ends, however, are stronger than the sides.

If you lay a hen's egg on its side, the narrow end also points ever so slightly downwards. Chicken breeders have long known that the orientation of an egg in an incubator affects its chances of hatching. Studies have shown that incubating eggs at different angles gives markedly different results – eggs positioned pointy end upwards fail to hatch more often than eggs positioned pointy end down. While the mechanism isn't fully understood, it seems that the wide end of the egg receives more oxygen than the pointy end, and is the location of the air sac, the pocket of oxygen that the developing chick inhales just before hatching. If the egg is accidentally placed wide end downwards, in an incubator, the head of the developing chick is at the opposite end from the air sac and will probably fail to make it out. The natural pear shape of the hen's egg may therefore be nature's way of increasing a chick's chances of survival.

Modern breeds of hens lay around an egg a day. Despite centuries of breeding and human interference, no one has managed to make a chicken lay any quicker. And while the world record for the most eggs laid in one year by a single hen is an impressive but sore-bottomed 371,* science has thankfully yet to squeeze any more out of hard-working hens. The gentle

* The highest authenticated rate of egg laying is 371 in 364 days, laid by a White Leghorn in 1979 at the College of Agriculture, University of Missouri, US (source: Guinness World Records – Most Prolific Chicken).

art of incubation, however, was one of the earliest of the hen's repertoire to be taken over by humans.

To hatch a hen's egg, it has to be kept constantly warm and cosy by its mother - the optimum temperature is around 37.5°C. Left to her own devices, a hen prefers to sit on a reassuringly large clutch of between eight and twelve eggs, rather than one egg at a time - the 'broody' hormone prolactin is thought only to be released only once she's laid a few. Once she's settled, she'll diligently cover her eggs for three weeks, only leaving for brief periods to grab food or have a drink. During the twenty-one-day period it takes for a chick to hatch, the hen will also turn her eggs regularly, using her beak to scoop under each egg and gently roll it over. This ensures that the developing chick doesn't get stuck to the shell membrane and the internal temperature is evenly distributed.

The embryo also needs a certain amount of humidity to grow properly - this moisture comes from the hen's own body and the surrounding environment. It's a delicate balancing act and one that doesn't always go to plan - about one in every five eggs naturally incubated by a mother hen will fail. It's astonishing, then, that any ancient civilisation managed to replicate this process. It's even more astonishing that they did it on an industrial scale.

The world has marvelled at many of Egypt's architectural wonders, but few have heard of their remarkable egg ovens. The eighteenth-century French scientist René Antoine Ferchault de Réaumur declared that 'Egypt ought to be prouder of them than her pyramids'. A bold claim, but this ancient culture's skill at egg-hatching was remarkable. Aristotle's *History of Animals*, written in 350 BC, makes the first mention of the country's attempts to artificially incubate eggs: 'In some cases, as in Egypt,' he wrote, 'they are hatched spontaneously in the ground, by being buried in dung heaps.'[3] Only two centuries later, Egyptian poultry farmers had got egg-incubating down to a fine art. Greek historian Diodorus Siculus was clearly impressed; in his *Library of History*, written in the first century BC, he enthuses: 'the men who have charge of poultry and geese, in addition to producing them in the natural way known to all mankind, raise them by their own hands, by virtue of a skill peculiar to them, in numbers beyond telling; for they do not use the birds for hatching the eggs, but, in effecting this themselves artificially by their own wit and skill in an astounding manner, they are not surpassed by the operations of nature'.[4]

The Egyptians called their ovens *Maamal-el-firak*' or 'the chickens' factory'. Réaumur was captivated by the idea of artificial incubation and had written extensively on the topic in his two-volume *The Art of Hatching and Rearing Domestic Birds of All Species in Any Season*. Originally published in Paris in 1749, a short abstract of his scholarly but long-winded work was translated by the English scientist Abraham Trembley and presented to the Royal Society a year later. The country's gentlemen of science were intrigued by the sheer scale of the Egyptian 'industry' and the closely guarded secrets of those who operated the ovens. 'The manner in which the eggs are hatched in Egypt', read the abstract, 'is well understood, only by the inhabitants of one single village, and those that live a short distance from it, about twenty leagues from Cairo in the Delta, which village is called Bermè. The Bermèans instruct their children in this art, and carefully conceal it from strangers.'[5]

The egg ovens, of which there were around four hundred at the time of Réaumur's book, were capable of holding between 40,000 and 80,000 eggs at a time. The ovens worked solidly for six months of the year, constantly rearing new batches of eggs for three weeks at a time. The process was paid for by a curious bargain - the owner of the egg oven only had to give back a number of live chicks equivalent to two-thirds of the total eggs he was given. If the operator of the oven was skilled at his job, he could hope to have a better incubation rate than this with as many as 80 or 90 per cent of eggs successfully hatching. Any surplus chicks were his to keep, eat or sell. The sheer number of poultry raised in this way was extraordinary, even by modern standards. Réaumur calculated that, even if only two-thirds of their eggs hatched, 'the ovens of Egypt give life yearly' to nearly thirteen million chickens.

The design of the Egyptian egg oven had developed over time. *The Travels of Sir John Mandeville*, a fourteenth-century travel memoir, described small furnaces in which the eggs were covered with horse dung to keep them warm; the manure slowly released heat as it decomposed, 'And at the end of three weeks or of a month [the people] come again', he wrote, 'and take their chickens and flourish them and bring them forth, so that all the country is full of them.'[6] In the early years of the sixteenth century, Sir Thomas More's *Utopia* imagined a fictional island where chickens were bountiful, an idea no doubt based on his knowledge of the Egyptian egg ovens. 'They breed an infinite multitude of chickens in a very curious manner,' he fantasised, 'for the hens do not sit and hatch them, but a vast number of eggs are laid in a gentle and equal heat in order to be hatched, and they are no sooner out of the shell, and able to stir about, but they seem to consider those that feed them as their mothers, and follow them as other chickens do the hen that hatched them.'[7] Later accounts, however, talked of brick-built structures more akin to bread ovens, in which fuel was gently burned to keep the eggs at a constant, and cosy, temperature.

Over the centuries, various European engineers and innovators tried to create their own versions of incubators. The sixteenth-century playwright and polymath Giambattista della Porta,* is thought to have drawn on his knowledge of ancient Egyptian models: 'But what I have done myself, and I have seen others do,' he explained in 1558, 'I shall briefly relate, that with little labour and without Hens, anyone may hatch eggs in a hot oven.' Della Porta made bold claims of hatching success rates that seemed almost impossible: 'If you do work diligently, as I have shewed you, in three hundred Eggs you shall hardly lose ten or twenty at most.' Careless boasting did della Porta no favours, however. His experiments and bold claims drew the attention of the Catholic Church's Inquisition, which was looking for signs of anyone meddling in the 'dark arts'.

* Giambattista della Porta is also well known for supposedly sneaking messages to close friends, imprisoned during the Inquisition, on the inside of hard-boiled eggs. Della Porta described an ancient recipe, grinding oak galls and alum with vinegar, to create an ink that soaks through the eggshell to leave a message on the inside of the egg but no trace of writing on the outside.

The almost magical success of della Porta's incubator quickly aroused suspicion, forcing him to stop any more dabbling in potentially heretical acts.[8]

The Dutch inventor Cornelis Drebbel, who is probably better known for building the first navigable submarine in the 1620s, had, years earlier, tried his hand at creating a self-regulating incubator. From the drawings, Drebbel's 'circulating oven' seemed to provide a way of automatically adjusting itself so that it could maintain a constant and ideal temperature for eggs. Writing over a hundred years later, the chronicler Cornelis van der Woude marvelled that Drebbel 'was able, by means of a strange and amusing device, to hatch duck and chicken eggs all the years round, yes, even in the middle of winter, without using ducks or chickens for this, and everything went so punctually, that the young were born at the proper time, just as if they had been hatched by ducks and hens'. As with many inventions of the day, however, Drebbel's incubator was viewed as a curiosity to amuse the royal court rather than a commercial breakthrough. Despite his brilliant mind and impressive achievements, Drebbel never made any money from his incubator or any other invention and ended his days impoverished and working in an alehouse.[9]

Fast-forward to the eighteenth century and Réaumur, keen to replicate the success of the Egyptians, also tried his own version of a 'manure incubator'. It was a culmination of over a year's worth of experiments conducted with the help of his able gardener and a yard of sprightly hens. After months of trial and error, and numerous disappointments, his green-fingered assistant, 'who had cared for so many unfortunate batches of eggs, and whose hope had been sustained like mine, came in one evening completely beside himself, to announce to me the good news we had anticipated for so long'. Réaumur could also barely contain his excitement. 'One of the eggs was crazed, that is to say, it had little cracks in its shell and the little chick that had made them could be heard from within.'[10]

At first, Réaumur's incubator was little more than a box of eggs buried in a dung heap, but he soon refined a design consisting of a wooden cask, filled with layers of shallow egg baskets, his own self-designed thermometer, and an air vent. A priest who ran a charitable commune in Paris asked Réaumur to come and inspect the commune's various buildings and grounds to see if there was space for one of his manure incubators. As Réaumur entered the priest's bakery, he noticed that an attic space above the huge ovens was perfectly balmy. 'I climbed up into this room; the heat I felt as soon as I entered made me suspect that this could be a stove as convenient as the Egyptian ovens for hatching chicks.' With a little direction from Réaumur, the commune's nuns set to work experimenting with the new hot-air egg oven and, by all accounts, achieved good results, hatching out a steady flow of fluffy chicks.

News of Réaumur's experiments spread. A French physician living in Canada wrote to Réaumur in the 1750s, thrilled that 'During the past year, and to even a greater extent during the present, people have eaten young chickens in February which had never before happened.'[11] After an initial flurry of excitement, however, interest in Réaumur's incubation techniques seemed to tail off, not least because many people had struggled to regulate the temperature of their incubators or achieve adequate ventilation for a decent hatching rate. Réaumur himself had noticed that the real magic of the Egyptian egg ovens lay not in their design but in the people who ran them. The care and attention lavished on the eggs made all the difference – every day the operator would lovingly turn all the eggs by hand and move them to cooler or warmer sections of the oven if needed. Thermometers were a new and not always reliable technology – Réaumur had even invented his own alcohol-based version to use in his incubator, to rival Daniel Gabriel Fahrenheit's mercury thermometer, but nothing beat the intuitive touch of the Bermèan hatcher, who could tell the temperature of the egg just by pressing it gently against his cheek or eyelid.

Prototypes for incubators continued to be developed throughout the nineteenth century. Mad, bad or simply doomed to failure, ideas ranged from heated Etruscan vases to forcing cockerels to act as mother hens. One unpleasant technique – called a 'living hatching machine' – involved trapping turkeys in small boxes, in a dark room, and making them sit on piles of chicken eggs. The turkeys weren't allowed to move, so were taken out once a day for force-feeding, and then plonked back on their clutch. To keep them calm, the birds were also 'given a glass of wine at dark, and an hour or two after chickens placed under her, which she would take to in the morning'.[12] Under such conditions, turkeys lasted only three to six months, but in the 1860s poultry breeding experts still viewed the scheme – from a commercial point of view – as a roaring success and 'the very best and cheapest way of hatching'.[13]Other notable and eccentric schemes included eggs being exposed to a bracing half hour of 'frictional electricity', incubators surrounded by heavy curtains for warmth and machines heated by natural hot springs.[14]

By the end of the century, a number of more reliable machines were on the market, including Hearson's oil-fired egg incubator, patented in England in 1881, and Cypher's self-ventilating incubator, which proved a hit on its release in the United States in 1896. Charles Cypher had also been tasked to create America's first mega-incubator by a Mr Truslow, a duck breeder from Pennsylvania. Truslow wanted to hatch on a vast Egyptian-like scale, rearing twenty thousand eggs or more at a time; Cypher not only designed but also built 'The Mammoth', a barn-sized incubator that worked a treat despite its vast size. Truslow, a man of few words, later wrote a glowing testimony: 'Gentlemen: I don't know what I can do better than give you some figures of this season's hatching. Set 22,848 eggs in Cyphers Incubators, and hatched 12,517 ducks, which is just short of 55 per cent, of all eggs set. Leaving off the first and last month of hatching, during which time the eggs are never good, set 17,250 eggs in Cyphers Incubators, and hatched 11,035 ducks, or just short of 64 per cent.'[15] Mass incubation had become a possibility.

What's interesting, perhaps, is just how ambivalent most farmers were about the idea. So many other areas of farming

– crops, livestock and fertiliser – had been revolutionised during the eighteenth and nineteenth centuries, but few viewed chickens, and their eggs, worth bothering with. Indeed, most farmers regarded the chicken coop as the preserve of the housewife, a footling pursuit that – at best – could provide a household with a few extra eggs for the family or a bit of pin money. Victorian agricultural improvers also seemed to view incubators as superfluous technology; as one writer noted in 1854, 'There was no adequate motive to pursue it in this country, where a quantity of poultry, fully equal, and even superior to the demand, may be raised by the natural means.'[16]

To understand this reticence, it's useful to know the historical laying habits of hens. When the agricultural writer Columella helpfully wrote a 'how to' for Roman chicken farmers, he jotted down a few notes about the kinds of bird worth keeping for eggs. While the pugnacious Greeks loved their cockfighting breeds, with tough, city-inspired Hellenic names such as the Tanagran, Rhodian or Chalcidian, Columella expressed a preference for prolific fowls with red or darkish plumage and black wings. White birds were best avoided, he insisted, 'too delicate and not very long-lived'; so too were bantams, 'unless one takes pleasure in their low stature'. The best egg layers were 'a red colour, square-built, big-breasted, with large heads, straight, red crests, white ears; they should be the largest obtainable which present this appearance and should not have an even number of claws. Those are reckoned the best-bred which have five toes.'[17] The chicken Columella describes sounds like a Dorking but we know very little about the breeds of fowl historical civilisations kept for egg production. Aristotle even talked of an Illyrian* hen that apparently laid 'thrice a day' – a claim that scholars don't believe a word of.

Few people realise that, until recently, eggs were seasonal produce. Hens would lay most of their eggs in springtime, to give their chicks the best chance of survival in warmer weather, with production tailing off slowly over summer and stopping altogether for a few months over winter. A number of factors affect how many eggs a hen will lay – how much daylight they get, how stressed they are, quality of diet and whether the chicken is moulting, which redirects her energy from laying to growing new feathers. Some of these elements have a greater effect than others.

Early farmers soon realised that some hens would carry on laying a bit longer into winter if they were given enough food and kept warm indoors. These 'everlasting layers', however, were few and far between, and most farmers couldn't afford to keep their poultry fed and cosy all year round. And even if they were able to do so, most hens were so stressed by confinement, or started developing health conditions, that they stopped laying anyway. If the world was going to change the egg-laying habits of the chicken, it was going to have to find a way to alter its biology.

In a fascinating study, a team of international scientists at the University of Oxford[18] tried to establish the moment in European history when people started to manipulate hens to

* In Classical antiquity, Illyria was in the north-western part of the Balkan Peninsula.

extend their seasonal egg-laying habits. And, more importantly, why. By analysing the ancient DNA in chicken remains, stretching back from the third century BC to the present day, researchers were able to detect alterations in the bird's genetic code. One gene, in particular, was revealing. TSHR or 'thyroid stimulating hormone receptor' plays an important role in hens' lives, controlling their growth, metabolic regulation and the way daylight affects their egg laying. Modern layers carry a modified version of this gene, which is associated with a quicker onset of egg laying at sexual maturity and a loss of strict seasonal reproduction; in other words, modern hens can lay later into the winter or even all year round. This modified gene also seems to make hens less stressed in large groups of other chickens or in confined settings. When the Oxford scientists established the time frame in which the altered TSHR gene became more common, they were surprised to discover a date of roughly AD 1000, which was much later than anyone expected.

Historically, however, the date makes perfect sense. Between the ninth and twelfth centuries, much of Europe underwent a shift in dietary behaviour. As we already know, fasting during Lent had been a Christian tradition since at least the fourth century. However, the idea of fasting days or 'lean days' became even more popular during the Middle Ages, as a way of people demonstrating their faith by giving up worldly pleasures. While the forty-day Lent fast required people to forgo all red meat, animal products and alcohol, many of the other fasting events only restricted the intake of red meat. Eggs and fish were still allowed. At its height, much of Europe was fasting at least one day a week – usually on a Friday, but sometimes on Wednesday or Saturday. Add these to further periods of abstinence during Advent and other religious festivals, and fasting days could number as many as 250 a year, leaving only a hundred or so days of normal eating. Eggs became more desirable than ever.

Along with pressures from religious edicts, increasing urbanisation may also have played its part. The Middle Ages saw a rise in both population and urban living; people settling in towns and cities were heavily reliant on food brought in from rural areas, but they also seem to have taken to buying and keeping their own backyard animals. Large beasts such as cows were impractical but small pigs, goats and flocks of chickens could be squeezed into cramped conditions and provide a reliable supply of milk, meat and eggs. Chickens that could tolerate the cramped conditions of a medieval urban backyard would have had a greater likelihood of living longer and raising more offspring. Medieval poultry keepers, who wanted birds that started laying at an earlier age and had superior egg-laying abilities, would also have favoured those hens that possessed the altered TSHR gene, a natural mutation; poor performers would have faced the chop. This kind of artificial selection over decades – in which birds with certain traits were more likely to survive and breed future generations – had a gradual but measurable effect on the chicken's seasonal laying habits and its temperament.

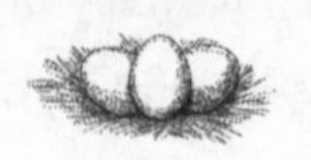

From man's earliest attempts to tame junglefowl to the beginning of the twentieth century, however, the average number of eggs people managed to wring out of a single hen rarely exceeded six dozen a year. Medieval egg tallies from manorial records, for example, show an average of around seventy to eighty eggs a year per bird.[19]

Even in the early twentieth century, some agricultural experts believed that getting a hen to lay more eggs than this simply wasn't possible; others thought that it was dangerous even to try.

As the writer and egg farmer Edgar Warren wrote in 1904: 'We hear a good deal said in these days about the 200 egg hen. Some are disposed to deny her existence, and to class her with such fabulous or semi-fabulous birds as the phoenix and dodo. Others admit that she has appeared in isolated instances, but is by no means common. Others contend that if she should appear in large numbers it would be a misfortune rather than otherwise, for such excessive egg production would weaken her system so that her eggs would not hatch healthy and vigorous chicks; and the 200 egg hen would be in constant danger of extinction from her own success.'[20] Given such scepticism, when, in 1913, American poultry specialist Professor James Dryden bred a hen that produced 303 eggs over a twelve-month period, people were rightly astonished. Dryden had been hired by Oregon Agricultural College in 1907 to head up their Poultry Husbandry department. He was on a mission: to breed a hen that would smash all existing records for egg production.

At the end of the eighteenth century, only one in every twenty Americans lived in cities. A hundred years later, it was one in four, and by 1920 the number of people living in urban areas matched those who lived in the countryside. This rapid growth in urban dwellers gave rural egg producers a boost, especially those located in the Midwestern corn belt, an area rich in grain for chicken feed and one that enjoyed easy rail access to the big cities. This burgeoning industry, however, still focused on a handful of pure breeds, convinced that these were the most prolific layers even though most couldn't produce more than a hundred eggs a year.

Dryden believed that a strong egg-laying ability was genetic but suspected breeders were looking in the wrong place. Many poultry farmers were still judging their birds by Victorian 'show' standards, wrongly assuming that there was a link between the fine appearance of a hen and its ability to be a good layer; they continued to favour pure-breed birds with attributes such as pleasing plumage, a symmetrical comb or smart, straight tail feathers. Speaking to one newspaper reporter in 1910, Dryden despaired that 'We think we are encouraging the poultry industry by paying premiums for feathers and other fancy points and for shape of body, and farmers go to the shows to purchase their breeding stock. They never suspect that the premiums indicate nothing of the egg-laying qualities of the fowl.'[21]

Dryden had a hunch that the real secret lay in mixed breeds or hybrids. These 'dunghill chickens', as they were often called, were the birds found free-ranging on small farms and backyards, with indeterminate parentage and a broad array of features.

Such birds, Dryden was certain, 'have better vitality, are more fertile, are less preyed upon by diseases and produce more eggs than the average flock of purebreds'. He was also convinced that poultry farmers should pay more attention to the specific laying habits of individual hens, only breeding from the superstars and sending the duffers to slaughter.

After a few attempts at crossing different breeds, Dryden successfully mated a White Leghorn, a bird originally from Italy, with a barred Plymouth Rock, a chicken breed popularised in the 1870s and itself a cross between black Java hens and a 'dunghill' cockerel with stripy plumage. The resulting White Leghorn-Barred Rock cross – marvellously named 'Lady MacDuff' – trounced the previous world record for egg laying, cracking through the 300-plus barrier for the first time in history. While the popular press trumpeted Dryden's, and Lady MacDuff's, achievements, the chicken-breeding world was less than thrilled.

As we learned in the last chapter, throughout the nineteenth century poulterers had grown rich promoting and selling pure breeds to farmers and fanciers alike. One of the most notable developments of the Agricultural Revolution of the eighteenth and early nineteenth centuries had been the 'improvement' of livestock, using the science of inbreeding among closely related animals to create better, fatter or more productive pure breeds. The kind of deliberate mixed breeding suggested by Dryden flew directly in the face of every agricultural proponent of 'High Farming' or, as the editor at Oregon's *Cottage Grove Leader* raged, 'This would mean an inevitable return in time to the razor-back hog and the inferior and mongrel breeds found a few decades ago in their native state before they were bred up to the present excellent standards by man.'[22]

Even more chillingly, the same newspaper compared Dryden's work to cross-breeding between human races, a state of affairs that went against 'Nature's law'. 'Of course,' he fumed, 'you might improve the characteristics and the qualifications of the Chinese or Africans by the infusion of the white race but it would be mighty hard on the Caucasians.'[23] The editor of the *Cottage Grove Leader* wasn't alone in his poisonous views. The eugenics movement in the US had gained increasing traction in the early years of the twentieth century, not least among those who were also interested in the genetics of animal breeding.

In his 1906 book *Inheritance in Poultry*, the chicken breeder and scientist Charles Davenport had no compunction about applying his theories about poultry genetics to the 'problem' of mixed-raced America: 'Again, if we accept the doctrine that man is a single species, all the momentous questions of human inheritance are questions of race inheritance. The outcome of such an admixture of races as is going on in America is a question of race inheritance. The offspring of a man and a woman having one or more diverse characteristics will follow the laws deduced from a study of crossed races. These are practical problems of human evolution, and experiments made with domesticated races can throw light upon them.'[24]

Davenport went on to create the Eugenics Record Office in 1910, whose primary goal was to 'improve' the human race by preventing unfit individuals from procreating; in reality, this meant the forced sterilisation of 'socially inadequate' people in state institutions, including the poor, the mentally ill and those in prison, among whom ethnic minorities were over-represented. Davenport's views hit fertile ground – America's early twentieth century was a period characterised by widespread economic unrest and rising rates of immigration. Overcrowded and rapidly expanding urban areas were increasingly struggling to deal with the social problems of crime and poverty; many government officials found it convenient to blame delinquency and unemployment on supposedly 'innate' characteristics of certain races, rather than on the wider problems of the urban poor. Davenport's eugenicist ideas of 'racial hygiene' not only appealed to some Americans but inspired a host of imitators across France, England, Sweden and other European countries including, of course, Germany. In the end, however, eugenics proved not only morally repugnant but also bad science. As Dryden's chicken experiments had shown, biological strength lay in genetic diversity, not furious inbreeding.

Dryden's book *Poultry Breeding and Management*, which he published in 1916, became the most popular and well-regarded textbook for ambitious egg farmers and remained in print for nearly thirty years. And although few farmers could reach Dryden's dizzy heights of over three hundred eggs a year, by the 1930s the average egg count had risen to a very respectable one hundred and eighty.[25] By increasing the number of eggs a hen could viably produce over twelve months, Dryden had removed one of the barriers faced by commercial egg producers. There were, however, plenty of other obstacles still to overcome.

Egg farmers were finding that even with the right cross-breed hen, laying still slowed down over the winter months. Their answer was to keep the hens indoors, under artificial lighting, but this soon created another problem – a lethal lack of vitamin D. Chickens, like humans, make vitamin D from sunlight; without it, they quickly become very poorly. Indeed, as late as 1916, one scientist complained that using chicks as experimental animals was close to impossible because of the 'many disorders and high mortality resulting when confined under laboratory conditions'.[26] Farmers who kept their chickens cooped up indoors – day after day – noticed that their birds experienced crippling leg weakness, poor plumage and a whole host of skeletal problems. Hens who were denied access to daylight also started to lay fewer eggs, with weaker shells and, if they were supposed to be laying fertilised eggs, were firing out more duds.

At that time, rickets was a common but little understood disease through the industrialised nations of the west; it had been known as a specific illness since its description by the English doctor Daniel Whistler in 1645 and had ravaged nineteenth-century cities with their poor nutrition, dark polluted skies and gloomy factories. Cod liver oil had been used medicinally, on and off, since Roman times as a cure for rickets without any scientific understanding of its properties. Only when nutritionists started to take a scientific approach to its study could its potential be fully realised.

In 1919, the British biochemist Edward Mellanby carried out an experiment feeding puppies on a diet of low-fat milk and bread to induce rickets. The young dogs, who were given yeast or orange juice, both touted as folk cures, still developed the disease, but when Mellanby added cod liver oil to their diet, the condition did not materialise. Something about cod liver oil was making all the difference. In the same year, the serendipitously named microbiologist Harriette Chick proved that cod liver oil and exposure to UV light could cure and prevent rickets. Mellanby and Chick – along with other experimenters – concluded that rickets was probably being caused by a lack of some crucial nutritional element. Building on the work of those who preceded him, in the early 1920s the American biochemist Elmer McCollum tried to establish whether the magic ingredient in cod liver oil was vitamin A or some new, and as yet unidentified, substance. He named his new discovery vitamin D.

Animal nutritionists soon put two and two together – indoor hens were suffering from a poultry version of rickets; by the mid-1920s, research studies were recommending that the vitamin found in cod liver oil was crucial if farmers wanted to avoid the problems suffered by chickens reared in unnatural confinement. It also became clear that chickens, in comparison with other animals that were studied – including rats, cattle and pigs – were acutely sensitive to a lack of vitamin D. If farmers wanted to move their flocks of hens indoors, they were going to have to do one of two things: give them cod liver oil or expose the birds to artificial UV radiation, which mimicked the effect of daylight. Only when one or both of these rules were followed could large-scale indoor egg production become a possibility. Whether it was beneficial for hens to spend their entire lives never once seeing the sky was a question yet to be raised.

One of the most bizarre outcomes of the introduction of indoor egg farming was the invention of chicken goggles. Hens confined in unnaturally tight spaces tend to peck aggressively at each other's feathers and, at times, will even try to eat each other. Gentle feather pecking is part of hens' normal preening and social behaviour; however, stressed, overcrowded and bored hens often redirect their foraging energy into pecking or cannibalising flock mates. The sight of a bloody injury on a fellow chicken can also be too much for another bird to resist pecking, often just out of idle curiosity. Many countries now debeak laying hens, a controversial practice developed in the 1930s that involves removing part of a day-old chick's beak,* but for the first half of the twentieth century, American poultry farmers had an unusual alternative.

Chicken goggles were just that – tiny eyeglasses designed to fit over the beak of the hen. Producers believed that the eyewear – which used rose-coloured lenses – would stop chickens being

* Debeaking is now banned in some European countries, although still legal in the UK. At the tip of the chicken's beak, tiny mechanoreceptors help it to finely discriminate what it's touching. Debeaked chickens often show pain-related behaviours such as tucking the bill under the wing or being reluctant to peck and preen.

able to distinguish red injuries or blood on other chickens and thus reduce the frequency of feather pecking. A number of different models were available - from Harry Potter-esque wire spectacles to red celluloid sunglasses - each fastened to the bird in different ways. A few models were completely opaque - more like horse blinkers - so the bird could only see sideways, rather than over its nostrils. Some gripped the beak like pince-nez, some were held in place by a strap or wire around the back of the head, while others - more distressingly - had split pins that were pushed through the chicken's septum.

Thanks to chirpy adverts from manufacturers such as the National Band & Tag Company - whose 'Anti-Pix' glasses will 'make a "sissy" of your toughest bird' - chicken spectacles sold like hot cakes throughout the US. The goggles even made it across the water to Britain, although fewer UK farmers were convinced of their efficacy. A Pathé newsreel from 1951 reveals their arrival on a farm in Spalding in Lincolnshire was deemed newsworthy. Glorious black-and-white footage shows an old poultry farmer patiently attaching chicken glasses onto one of his hens, who 'talked too much and got pecked for her pains'. 'Lucky for her,' the newsreader quipped, 'Farmer Atkinson found the one cure for hen pecks - rose coloured spectacles which clip on the beak. With glamour like this to soothe her hurt feelings, even a Rhode Island Red can take a fresh view.'[27]

While chicken goggles didn't catch on long term, another egg-inspired invention did. One of the biggest challenges for egg producers was getting their precious cargo to market. As we already know, eggs are most vulnerable on their sides - carrying eggs of any quantity in a basket or loose in a box often resulted in disaster, especially given the poor state of rural roads until well into the twentieth century. Canadian newspaper editor and part-time inventor Joseph Coyle famously overheard an argument between a local farmer and the hotel owner to whom he was attempting to deliver eggs. The hotelier was furious that so many of the eggs he'd bought had arrived smashed. Coyle quickly set to work and in 1911 revealed his new idea – a paper carton with individually paper-cushioned slots into which eggs could be placed standing up. What started with a prototype that needed to be folded by hand soon became so popular that Coyle was forced to come up with a new machine that could manufacture his 'Coyle Egg-Safety Carton' by the dozen.

Given that rates of egg consumption in America had rocketed in the early years of the twentieth century, to more than one egg per person per day,* Coyle wasn't the only one turning his or her attention to egg boxes; just two years later, Virginia-based inventor Stuart Ellis devised a metal box filled with card cylinders, like toilet-roll inners, which kept each egg apart for long-distance journeys, while back in England Thomas Peter Bethell patented the Raylite Egg Box, a series of interlocking cardboard strips that separated and protected the eggs in transport.

In the end, neither Coyle, Bethell nor Ellis's designs could compete with the brilliantly simple and economical egg box

* According to the US Department of Agriculture, Americans reached peak egg consumption in 1945, averaging 404 per person in just one year. By the 1990s it was just over half that figure. (Source: *Washington Post*, 28 Feb 2019)

designed by Francis Sherman of Massachusetts in 1931. Sherman had the brainwave to use pulp-fibre, a cheap fibrous material that could be moulded into endless rows of boxes. His second stroke of genius was to produce the egg cartons in a continuous run, from which both the top and bottom of the box could be cut in one piece and then folded in half to create a hinge. The design of modern egg boxes remains virtually unchanged.

Egg boxes often come in sixes or twelves. But why? There are some tantalising possibilities, but no one really knows. One thought is that, since ancient times, twelve has held magical significance thanks to the twelve lunar cycles in a year. The ancient Mesopotamians and Egyptians both used duo-decimal counting for many aspects of their lives, including trading goods - humans have twelve finger bones on one hand, three on each finger, so it was easy to count up to twelve on one hand, with the thumb acting as the pointer. Another possibility is that twelve is popular because it is such a useful number when it comes to division. Unlike other numbers near it, such as ten or eleven, the number twelve is the smallest number to contain six factors - 1, 2, 3, 4, 6 and 12 - making it an easy number to divide and subdivide. The most likely answer, however, relates to the pre-decimal system of money that was common in Britain and her colonies. With 4 farthings in a penny, 12 pence in a shilling and 240 pence in a pound, dealing with small items such as eggs in dozens made mental maths a good deal more straightforward.

Perhaps most interesting of all, however, is that eggs were very nearly not 'eggs' at all. Fifteenth-century England was a melting pot of different regional dialects, often with little agreement on spelling, pronunciation and even which words to use for everyday objects. This presented a headache for the printer William Caxton when he tried to translate Virgil's *Aeneid* in 1490, a conundrum he perfectly illustrated in the preface to his book. In it, Caxton told the story of a heated conversation between a merchant from the north of England and an egg seller from the south. While the northerner used the word *egges*, from the Old Norse, the southerner insisted on calling them *eyren,* the even more ancient Anglo-Saxon term. The egg seller then snapped at the merchant, misunderstanding his dialect, insisting that he could 'speke no frenshe'. Caxton despaired at the challenge of reconciling the hotchpotch of competing tongues; 'what sholde a man in thyse dayes now wryte, egges or eyren, certainly it is harde to playse every man'.[28] For years, the two words sat side by side, jostling for supremacy, and the Norse 'egg' didn't emerge as the victor until well into the sixteenth century. Pockets of southern England, however, hung on tenaciously; *eiren* was still being used in Kent dialect in the nineteenth century before dropping out of use altogether.

7

BROILERS

Oven Ready

Plymouth Rock

People have been eating chicken for a very long time. Ancient Rome's most celebrated cookbook - *De Re Coquinaria* (On the Subject of Cooking) - devoted an entire chapter to the bird, including crowd-pleasing dishes such as 'Chicken Parthian Style' (*Pullum Parthicum*), a one-pot roast with pepper, lovage and wine, and 'Chicken with Laser' (*Pullum laseratum*),[1] a mysterious herb now thought to be extinct. Chicken featured in historical recipes from northern Europe too. The thirteenth-century manual *Libellus De Arte Coquinaria* (The Little Book of Culinary Arts) contained dozens of recipes written in Danish, Icelandic and Low German. Many of these meals seem reassuringly familiar; *kloten en honœr* or 'chicken dumplings', for example, cooked in wine, herbs and bacon fat, wouldn't be out of place on a modern Scandinavian menu.[2]

From these and other ancient cookbooks, it's clear that chicken meat was a familiar ingredient. What's more difficult to establish, however, is just how many people ate chicken with any regularity. Many of the dishes in these surviving recipes are clearly aimed at the wealthy - the inclusion of expensive spices such as cumin suggests they weren't for ordinary folk. Recipe books were also written for the few who were literate and lucky enough to have access to the written word. It's perhaps telling that at King Henry IV's coronation banquet in 1399, chicken was deemed a food glamorous and indulgent enough for the king's table. Along with a boar's head, a baby swan, herons, suckling pigs, peacocks and cranes, chickens were a key part of the sumptuous feast.

In Britain today, roughly 90 per cent of the population eat meat and most eat chicken at least twice a week.[3] But how long has chicken meat been such a widely consumed food? While we know that the Romans became adept at farming chickens, we don't know the relative proportions of birds raised for eggs and those for meat. Many archaeologists suspect that chicken was a delicacy until relatively recently, rather than an everyday meal, and that chicken meat was largely a by-product of egg production, with its superfluous cockerels and spent hens. There were also certainly differences between the historic diets of the rich and the poor, and those living either in dense urban areas or remote backwaters, factors that aren't always revealed in either the written or the archaeological records. With all this uncertainty, what meaningful things can we say about the historical consumption of chicken meat?

The first thing we can say is that, until recently, most chickens were smaller than the birds we find on our supermarket shelves today. Archaeologists have compared the tibia lengths of ancient chicken bones, from historic sites across London, and modern bones; leg length is a good proxy for overall body size, and it seems the average skeleton size of a chicken remained remarkably consistent for hundreds of years. From Roman times to the fourteenth century, the average chicken was not much larger than its wild cousin, the junglefowl. Between the fourteenth century and the mid-nineteenth century, the average length of chickens' leg bones seemed to get fractionally longer, suggesting slightly taller birds, but for all intents and purposes most chickens until the mid-1800s were lucky to weigh more than a kilogram – about the same as a pheasant.

By contrast, chicken breeders in the Far East had been perfecting larger birds for thousands of years. Ancient breeds such as the Malay or the Indian Aseel, with their long necks and upright stance, would have towered over the average Western chicken, but their purpose was primarily for fighting, not for eating. The best cockfighting birds needed muscular power, not fat, and so many of these centuries-old breeds were strong, lean birds, making them fierce fighters but tough dinners. Even when exotic game birds such as Queen Victoria's Cochin-Chinas started to appear on English soil, they proved too feisty to farm for either eggs or meat. The 'natural pugnacity of their disposition, which shows itself at the earliest possible period', despaired one contemporary writer, 'deters most breeders, excepting those who breed for the cock-pit. I have many times had whole broods, scarcely feathered, stone-blind from fighting, to the very smallest individuals; these rival couples moping in corners, and renewing their battles on obtaining the first ray of light.'[4]

Some of the richest and most interesting information we have on chicken consumption in Britain comes from the Middle Ages, in the form of manorial records that carefully documented the kinds of animals being raised on medieval estates. What's striking is just how few landholders bothered with more than a handful of chickens. From the middle of the thirteenth to the beginning of the fifteenth century, only a tiny proportion – just 2 per cent – of all the recorded livestock was poultry. While most estate farms had a small flock of chickens – between around five and fifteen birds – only a handful of larger manors had more than fifty and nearly a fifth of estates had no chickens at all.[5] And although hens' eggs feature heavily in medieval recipes, the amount of chicken being consumed in wealthy households was less than 10 per cent of all the meat consumed.[6]

One of the reasons for this may have been the sheer variety of wild and domestic birds available to the lord of the manor and his family. *The Forme of Cury,*[*] an extensive collection of fourteenth-century recipes thought to have been compiled by Richard II's chefs, includes lists of birds that read like a twitcher's handbook. Alongside chickens, many other fowl made it into the pot, including geese, pigeons, ducks, pheasants, partridges, quail, swans and their cygnets, and even curlews.[7] One of the earliest compilations of recipes from the Middle Ages, *Le Viandier*, includes instructions for how to rustle up blackbird and

* The title *The Forme of Cury* means 'The Method of Cooking', with 'cury' coming from the Middle French *cuire* 'to cook' and not 'curry', as in the Indian dish, which comes from the Tamil *kari*.

thrush pie, and woodpecker pasty.[8] With such a smorgasbord of available avians, it's perhaps no surprise that chicken meat wasn't always the go-to white meat for wealthy households.

There are, however, records of medieval feasts and household inventories that do list chicken as a meat, but often in the form of 'capons' or castrated cockerels. On the medieval manor farm, this served a number of useful functions: not only were neutered cockerels less aggressive, which meant they could happily live together in a flock with other hens and castrated cockerels, but the act of removing a bird's testes also turned it into a docile eating machine. Capons were prized for their ability to pile on the pounds - a mature capon would weigh more, and have more buttery, fatty meat, than either a hen or a sexually mature cockerel of the same age. Capons also seemed to lack the stringiness or gamy taste of cockerels - a by-product of their sedentary, overfed existence.

And capons conferred other benefits. While small but tender 'spring chickens' were usually eaten no later than summer, at just a few months of age, a capon took nine months to a year to mature, making it ready for consumption in winter or very early spring, when almost no other fresh, tender meat was available. Capons could also be used as surrogate hens, as chick mothers and egg incubators. Writing in the sixteenth century, the natural historian Conrad Gessner noted the unusual way farmers encouraged their capons to adopt a clutch: 'make him drunk with wine-soaked bread and take him quickly to a dark place. As soon as he regains consciousness, he thinks the eggs belong to him and incubates them.'[9]

You'd be forgiven for wondering how a bird can be castrated. The word 'capon' comes from the Latin *caponem*, itself from the older root *skep* or *kep*, meaning to cut or scrape,* but, as we already learned in Chapter 1, cockerels have their genitals on the inside of their bodies. Castration or caponisation, therefore, was an unpleasant and very tricky procedure, fraught with risk. The cockerels would often be operated on around five months of age – just before they reached sexual maturity – and it was a job for a skilled 'caponiser', a job title now thankfully obsolete.

As with many rituals of husbandry from early history, the timing of caponisation was related to feast days and other religious events in the calendar. The time for the cockerels' 'snip' usually fell between the Feast of the Assumption (15 August) and the Nativity of Mary (8 September). When that three-week period came around, each unfortunate chicken was lashed to a table or the top of a barrel and an incision was made between two ribs, without anaesthetic. Once the hole was made and opened up, the caponiser would search for one of the internal testes – which in an immature cockerel could be as small as a rice grain – before yanking, slicing or snipping it out. The poor cockerel would then be flipped over onto its other side and the entire ordeal repeated. Needless to say, with such a barbaric and unsanitary procedure, the fatality rate was high. One record from Sedgeford, in Norfolk, in 1375 shows that out of an attempt to caponise eighty-two cockerels, thirty died in the process.[10] This figure is unusually high and probably reflects the efforts

* 'Scab' comes from the same root.

of a cack-handed amateur; but the average mortality, even with skilled practitioners, was still about one in every seven birds.[11]

If the poor capon survived its emasculation, it could look forward to a few months of respite before its final humiliation. 'Cramming' is now, thankfully, a largely forgotten trick of the chicken farmer, but continued to be enthusiastically carried out well into the twentieth century. The practice was an ancient one; Pliny the Elder mentions it in his *Natural History*, in the first century, believing it to have originated on the Greek island of Delos, a wealthy commercial and cultural centre that thrived in the second and first century BC. Unusually for the time, Pliny seemed troubled by its excesses and cruelty: 'The people of Delos were the first to cram poultry; and it is with them that originated that abominable mania for devouring fattened birds, larded with the grease of their own bodies.'[12]

Crammers or 'higglers', as they were sometimes known, were employed to force-feed capons for the last two or three weeks before they were slaughtered. For those who wanted to cram their own birds, top tips were plentiful. The professional chef Robert May, who trained in France and worked in England in the seventeenth century, wrote down the technique for his fellow gastronomes to follow. *The Accomplisht Cook, or the Art and Mystery of Cooking* described the process in a chapter cheerfully named 'Excellent Wayes for Feeding of Poultrey': 'The best way to cram a capon (setting all strange inventions apart) is to take barley-meal reasonably sifted, and mixing it with new milk, make it into a good stiff dough; then make it into long crams thickest in the middle, and small at both ends; then wetting them in lukewarm milk, give the capons a full gorge thereof three times a day, morning, noon and night, and he will in a fortnight or three weeks be as fat as any man need to eat.'[13]

It's interesting that May mentions other 'strange inventions'; the practicality of stuffing a sausage of wet dough, by hand, down the throat of a reluctant chicken must have been a challenge. Unsurprisingly, cramming the machine soon took over, a simple hopper on a stand that, with the press of a pedal or hand pump, released a set amount of food, via a tube, directly down the chicken's throat and into its crop (a kind of pre-stomach storage area). Like the operator of a teddy-bear stuffing machine, the crammer would grab a bird, push its beak and throat onto the horizontally mounted feed tube, and then press a quantity of food into its body. Repeating the process two or three times a day, for two or three weeks, artificially inflated the poor capon to almost twice its original weight.

Taking into account the high mortality rates, the cost of feeding a capon extra food and the labour involved in looking after and castrating cockerels, it's no wonder that capons were almost exclusively the food of the rich. In 1213, King John ordered 3,000 capons, 1,000 salted eels, 400 pigs, 100 lb of almonds and 24 casks of wine for his Christmas feast. Coronations were also celebrated with capons, often in a spiced stew called *dillegrout*. The porridge-like dish was made from almond milk, spices and capons and was first presented in 1068 at the coronation of

Matilda of Flanders, wife to William the Conqueror. It proved such a hit that the same dish was still being served at the coronation of George IV in 1821.

At a royal banquet, however, perhaps no dish was more impressive than a *cokagrys* or *cockentrice*.* Guests could enjoy the bizarre sight of a gruesome hybrid roast – half capon, half suckling pig, sewn together to create an edible chimera. According to one late fifteenth-century recipe, to make the fantastical beast, a cook must 'take a capon, scald it, drain it clean, then cut it in half at the waist; take a pig, scald it, drain it as the capon, and also cut it in half at the waist; take needle and thread and sew the front part of the capon to the back part of the pig; and the front part of the pig to the back part of the capon, and then stuff it as you would stuff a pig; put it on a spit, and roast it: and when it is done, gild it on the outside with egg yolks, ginger, saffron, and parsley juice; and then serve it forth for a royal meat.'[14] If that failed to amuse, you could always serve a 'Helmeted Cock', a cooked capon dressed in military regalia, straddling a suckling pig as if riding into battle.[15]

A hundred years later, capons were still an indulgence few could afford. In the play *As You Like It* (*c.*1600), when Shakespeare describes a middle-aged judge at his peak of material success, he can think of no better allusion than 'the justice, In fair round belly, with a good capon lin'd'.[16] As with so many of Shakespeare's exquisitely turned phrases, the

* Not to be confused with the *cockatrice*, the terrifying mythical half cockerel, half serpent from Chapter 3.

words also had other meanings now lost on modern audiences but ripe for Elizabethan crowds.* 'Capon' was also a crude insult, meaning unmanly and dull, while a 'capon justice' was a corrupt magistrate who was bribed with gifts of expensive food, especially capons.[17]

While capons graced the tables of the wealthy, we know much less about the everyday food of the medieval peasant. We can guess, however, that as with other domesticated animals such as sheep, ordinary people would have eaten chicken only when a hen had exhausted its egg laying or a cockerel was superfluous to requirements. A good layer, a good breeder or a good brooder was too important to eat. It's telling that many peasant recipes from before the twentieth century involve plenty of long, slow cooking time in liquid – the only way to make tough older flesh palatable. Dorothy Hartley's *Food in England,* a wonderful exploration of historical food, finds recipes such as 'Old Fowl for Pot-Roast', Cockaleekie soup – which is apparently best made with 'the oldest cock' – and the rather alarming-sounding 'A Very Tough Old Fowl Exploded'.[18]

One of the only ways we can peek into the world of medieval peasant diets is by looking at food residues left on ancient sherds of pottery. Archaeological excavations tell us that, for the everyday person, the main cooking facility would have been a pottery 'stewpot', which would have sat in or close to the hot

* Shakespeare also used the word 'capon' to denote a love letter, a meaning that used to flummox scholars. It turns out that folding letters in certain ways had different social meanings in this period – the 'capon' letter was folded to look like it had two wings, like a chicken, and would have instantly told the recipient that its contents were amorous and secret.

embers of a fire. Microscopic traces of fats and other residues on these pots reveal the dominance of potage and thick soup as well as dairy products in the diet of everyday folk. Grains were eaten either in the form of bread or in slow-cooked stews, flavoured with small amounts of meat or meat fat. Leeks, brassicas and herbs were common home-grown ingredients and, while chicken eggs were eaten, evidence of poultry flesh is more difficult to find.

Of the meat consumed, it seems mutton, pork, beef and even horse[19] were more likely to find their way into the stewpot than chicken, but only once the relevant animal's main function - such as producing milk, wool or draught power - had been exhausted. 'Piers Plowman', a poem written at the end of the fourteenth century, perfectly demonstrated the privations of the average peasant's diet:

> *'I have no penny,' quoth Piers, 'pullets for to buy,*
> *Nor neither geese nor pigs but two green [unripe] cheeses,*
> *A few curds and cream and an oaten cake,*
> *And two loaves of beans and bran baked for my children.*
> *And yet I say, by my soul, I have no salt bacon;*
> *Nor no hen's eggs, by Christ, collops* for to make.*
> *But I have parsley and leeks with many cabbages.'*[20]

* A collop was a slice of bacon or an egg fried with bacon.

Little had changed by the Georgian era. The fabulously named Bonington Moubray Esquire, writing in the early part of the nineteenth century, noticed his fellow Englishmen's reluctance to either farm or eat chicken in any great quantity. 'In Britain,' he noted, 'where a greater quantity of butcher's-meat is consumed than probably in any other part of the world, poultry has ever been deemed a luxury, and consequently not reared in such considerable quantities as in France, Egypt, and some other countries, where it is used more as a necessary article of food, than as a delicacy for the sick, or a luxury for the table.'[21] Over on the other side of the Atlantic, a similar picture had emerged – even as late as the 1920s, most Americans still viewed chicken as an aspirational treat. In 1928 the Republican Party, in a now famous newspaper campaign, claimed it was not a 'poor man's party' but a political force for wealth and opportunity; 'Republican prosperity has reduced hours and increased earning capacity,' they boasted, 'silenced discontent, put the proverbial "chicken in every pot."'

The choice of a poultry-themed boast was an interesting one. The quote is thought to have originated in a declaration made by Henri de Bourbon, who ruled as King Henri IV of France from 1589 to 1610 and made the commitment: 'If God gives me more life I will make sure that no peasant in my realm will lack the means to have a chicken in his pot.'[22] The claim made by the US political campaign over three hundred years later was almost identical to the French king's promise; it demonstrated the bird's value – chicken was the same price as rump steak and even more expensive than either pork or ham.[23] And while it's doubtful whether the Republicans did, indeed, put a chicken dinner within reach of every household, the claim was clearly designed to impress.

A few years earlier, America's Bureau of Foreign Commerce had released a report that highlighted the problems and potential of chicken farming as a money-making business. The author, George Tanner, a diplomat stationed in Belgium, drew attention to the staggering amounts of money being made by the French in particular, but also in Holland and Belgium, all three of whom were exporting an enormous quantity of eggs to England. According to the report, the trend only seemed to be going one way. Between 1856 and 1874 the value of eggs exported to England from French farms alone had multiplied by a factor of fourteen from around $1 million worth a year to over $14 million (over $300 million in today's money).

This extraordinary feat of commerce, which still left enough eggs for France's own population to enjoy, was achieved not through huge, well-organised poultry enterprises, but a network of numerous small independent sellers. These were often women running a side-line on the family's farm who sold their handfuls of eggs to middlemen, who then exported them to England. The profits were small for the individuals selling the eggs, but the volume of trade was vast. Even faced with the challenges of high land rents, the costs of building accommodation for the birds, feed and caretakers for the chickens, the inevitable losses from disease and the bitter cold of the French winter, the egg producers still felt it was worthwhile.

Why wasn't America, the land of opportunity, getting a slice of the action, asked the report? 'Poultry thrive nowhere so well as they do in the United States', argued Tanner, not only due to a more favourable climate but also because practically everything was cheaper for the potential farmer – construction timber, grain, availability of farmland. The same logic, he urged, could be applied to raising chickens for meat. 'Those who know the European climate, know how unfavourable it is, know the expenses that attach to raising poultry here, can at once see the immense advantage the American poulterer has over the European. He has this advantage in everything if it could be but followed up. His fowls, which can be "raised" with little or no effort on his part, can be made comparatively as great a source of revenue to him as the hog, or his wheat or cotton.'[24]

The report made a plea to the men of America to rise to the challenge of chicken farming, but it was actually a quiet farmer's wife who is celebrated – or blamed, depending on your perspective – for founding the 'broiler' industry as we know it today. (Broiler is a name given to any chicken raised for meat but comes from the cooking term 'to broil', which means to grill or cook with direct heat.) Cecile and David Wilmer Steele were poor farmers living in a rural backwater on the Delmarva Peninsula in Delaware. David had managed to land a job with the local coastguard, leaving Cecile to manage the farm by herself, selling eggs from their small flock of hens.

In 1923, to replenish the flock, Cecile ordered fifty fertile eggs from the local hatchery. Owning to a clerical error, the hatchery mistakenly sent 500 eggs. Landed with ten times more fertile eggs than she'd asked for, Cecile kept the error to herself, built a large shed and decided to see if she could raise and fatten the birds to sell as meat, as she couldn't afford to keep them all as long-term layers. After just eighteen weeks, and a generous feeding regime, three-quarters of the birds had survived and reached two pounds in weight, then viewed as a suitable size for market. Most importantly, Cecile had made a tidy profit and a quick return. The following year she bought 1,000 eggs, the next year after that 10,000. Other local farmers watched Cecile's success with interest and soon started their own broiler flocks. By 1928, Cecile was raising 26,000 broiler chickens, with another five hundred farmers on the Delmarva Peninsula hatching their own broiler operations, raising chicken for some of the largest cities on America's east coast.

It was a deceptively straightforward business model and one that a handful of other farmers across America had attempted as early as the 1880s but with no real lasting success. Cecile's enterprise, on the other hand, had hit a perfect storm of circumstances both local and national. The 'Roaring Twenties' was a period of dramatic economic growth for the US; the nation's total wealth more than doubled during the decade, while the average income per head rose by over a quarter. Commerce was riding high thanks to a heady combination of abundant raw materials, especially coal, iron and oil, new techniques of mass production, and tariffs that protected industry from foreign competition. Many urban families were finding they could afford to spend a little more on 'luxury' foods, including the occasional chicken. At the same time, traditional sectors of American farming were experiencing a catastrophic nosedive; the mechanisation of agriculture and overproduction of crops had led to a slump in grain prices – wholesale wheat plummeted from a high of $2.45 per bushel in 1920 to just 49 cents in 1932.[25] For farmers thinking about switching to broilers, grain was suddenly more affordable than it had ever been.

As we already know, by the early twentieth century, chicken farmers in both Britain and America had found a solution to the problem of mass incubation of eggs. This was crucial as it outsourced the time-consuming business of breeding new stock to hatcheries. Poultry keepers like Cecile, rather than having to maintain their own flock of breeding hens, could order scores of fertile eggs or hatched chicks to be delivered directly to their doorstep. At the same time, improvements in transportation and refrigeration, the rise of large grocery store chains and the willingness of hatcheries and feed companies to provide easy credit to broiler farmers all boosted business. Egg producers in the Delmarva Peninsula were also having a particularly bad time in the mid-1920s, with many of their birds succumbing to an outbreak of Marek's disease, which causes paralysis in chickens. The breed they farmed was the White Leghorn, a bird chosen for its ability to lay consistently. It was also unusually susceptible to Marek's but, critically, only after the bird reached three months of age, by which point it could be fattened up and sold as a broiler instead. Switching to broilers offered desperate egg producers a lifeline.

The Delmarva Peninsula, on the eastern seaboard, was also perfectly placed to send chickens to some of America's biggest cities – New York, Washington, DC, and Philadelphia. New York, in particular, was a market eager for chicken meat – by the early twentieth century, it had become the largest Jewish city in the world and the place where almost 75 per cent of all America's first- and second-generation immigrant Jewish families had made their home. Not permitted to eat pork by kosher dietary rules, two million Jewish families were responsible for buying over four-fifths of all the chicken in New York in the 1920s. Chicken was not only attractive because it fitted with religious custom but, being over twice as expensive as other meat, it was also seen as suitably special to serve on the sabbath and Jewish holidays. Delmarva farmers such as Cecile could send their birds, still alive and squawking, directly to New York, where brokers would buy the chickens and distribute them to kosher butchers. For over a decade, the birds were transported live, but by the late 1930s the first processing plants opened in Delaware that could slaughter and ship ice-packed birds straight to urban centres to an increasingly non-Jewish market and a less strict second-generation Jewish community.

When the United States entered the Second World War, Americans were eating more chicken than they had ever done but it still wasn't an everyday staple. During the conflict, the US government rationed red meat but chicken remained freely available, giving the broiler industry an extra boost from consumer demand. In 1939, the *American Poultry* journal asked experts the question 'How Will the War Affect the U.S. Poultry

Industry?' One contributor remarked, 'chickens are not a war commodity... Armies don't eat chicken',[26] but only a year later, broiler birds – especially from the Delmarva region – were being requisitioned to feed soldiers as part of the war effort, putting an even greater pressure on production facilities. Rearing chickens for meat was still a hugely labour-intensive process. Providing the basics for poultry – the accommodation, the heat from coal stoves, the endless sacks of feed, the water and the daily care – took physical effort. Even in death, the chickens needed complex handling – birds had to be slaughtered, processed and transported, all tasks that usually fell to the country's young, fit, male workforce. The only problem was that many of them had left the country to fight. And so began one of the most extraordinary and little-known stories in the fowl's history: the use of German prisoners to keep chicken on the US Army's dinner plate.

After three years at war, Britain was fast running out of space to house the thousands of prisoners of war it had captured during the conflict, and called on America for help. Between 1942 and the end of the war, more than four hundred thousand enemy POWs were sent to the United States and housed in rural camps across the country. There was a particularly acute labour shortage for the Delmarva broiler industry – not only were many of its men fighting abroad but plenty of female workers who stayed behind moved from processing plants to canneries, which paid higher wages. The owners of six chicken-processing factories in Delmarva decided to join forces to build a POW camp and by the summer of 1944, the first three hundred captured

German soldiers started work. By the end of the war, over three thousand POWs were involved in keeping the peninsula's hatcheries, farms, processing plants and feed mills open, mostly to feed demand from the US Army.[27]

By 1945, people's attitude to chicken meat had fundamentally changed but the industry needed to reduce costs and improve efficiency if white meat was ever going to be an affordable, everyday food. Despite the fact that factories were beginning to adopt labour-saving devices and conveyor-like production systems, the end result was still hampered by the fact that the average chicken weighed less than a bag of sugar. The secret to making the broiler business truly profitable lay in one place – the creation of a 'mega-bird'. To achieve this, the industry needed two things. The first was to find a breed or cross-breed of chicken that was naturally bigger and meatier than its cousins. The second was to come up with a way to make those chickens grow large using as little feed as possible. And if they could grow quickly, even better.

As we already know, the average size of a chicken destined for the table had remained fairly constant from Roman times, through the medieval era and into the nineteenth century, and weighed only a little more than a kilogram.[28] Cramming a capon for the last few weeks of its miserable life certainly made it more corpulent but it didn't change the fundamental genetics of a breed. A number of very large chickens had been introduced to

Britain and America during the Fowl Mania craze of the mid-nineteenth century but they didn't necessarily make good eating or suitable candidates for farming. Malays, like those that were gifted to Queen Victoria, were often too feisty to raise together in a large flock and only thrived in hot weather, while others, such as the fluffy, teddy-bear-like Cochins and Brahmas, were docile and hardy but not particularly good layers (and therefore difficult to hatch in any volume), slow to grow to full size and produced a coarse, dark meat not to everyone's taste.

In 1945, however, the Great Atlantic & Pacific Tea Company decided to launch a competition, in partnership with the US Department of Agriculture, to find the world's best meat chicken breed. With no shortage of ambition, the competition was called 'The Chicken of Tomorrow' contest and its aim was to find the holy grail of poultry – 'A broad-breasted bird with bigger drumsticks, plumper thighs and layers of white meat.'[29] Like a meaty beauty pageant, contestants were invited to take part in a series of state and regional heats that climaxed in a national final. Thousands of dollars in cash prizes were up for grabs for the winners in various categories such as 'Best Skin Texture' and 'Uniformity of Size'. Farmers and breeders from all over the country were invited to take part, each sending in a selection of fertile eggs to special competition hatcheries, where their chicks would be hatched under strictly controlled conditions. From there, each contestant's chicks were raised under exactly the same conditions, with their eating habits and weight gain scrupulously measured and documented. This was a competition that wasn't just looking for the meatiest bird, but a chicken that had the best 'feed efficiency', the creature that could most easily turn its food into flesh.

After twelve weeks, the chicken entrants were live-weighed, killed and their carcasses analysed for appearance, ratio of white to dark meat and other fleshy qualities. From the forty finalists, the national winner for 'Carcass Characteristics' was the meaty, white-feathered 'Arbor Acres' Plymouth Rock entered by Henry Saglio. Saglio was a chicken farmer from Connecticut who had started breeding white-feathered birds because he observed that stray black feathers left on after plucking were more noticeable than pale ones. The victors in the 'Economy of Production' category were brothers Charles and Kenneth Vantress, owners of the Vantress Hatchery. The brothers had crossed two types of chicken: a male Cornish game bird, a short stocky cockerel with a broad breast but too slow-growing to be a commercial meat bird, and a New Hampshire Red, a docile, greedy hen blessed with early maturity and easy weight gain. The resulting 'Cornish Cross' smashed the competition for feed efficiency and weight.

These two different types of chicken – the Cornish Cross and the White Plymouth Rock – were to become the foundation for the modern broiler industry. The Chicken of Tomorrow contest uncovered the winning combination – almost all modern broilers are based on one or both of these cross-breed birds. Since the 1950s, chicken genetics companies have refined the breeds even further, using intensive and sophisticated genetic selection to create a broiler that now bears little relation to its prize-winning predecessors. In the 1920s, it took a farmer over a hundred days and a costly 5 kg (12 lb) of grain to raise one meat chicken, which would, at slaughter, weigh just over 1 kg (2.5 lb). After the Chicken of Tomorrow contest, the time it took for a bird to reach slaughter weight began to fall year on year, accompanied by a concomitant rise in body size and reduction in the amount of food needed – by the mid-1950s it took seventy days for a chicken to reach 1.3 kg (3 lb) in weight eating just 4 kg (9 lb) of grain; by the mid 1970s, the same type of bird grew to a new record of 1.7 kg (3.8 lb) in fifty-six days, eating just 2.7 kg (6 lb) of food.

Since then, the trend has continued apace and thanks to decades of research and the development of fast-growing breeds – bolstered by high-energy feed, added vitamins and the routine use of antibiotics and other medications – broiler chickens are now behemoths compared to their former selves. In 2020, according to the US National Chicken Council, the average broiler chicken weighed nearly three times as much as its early twentieth-century predecessor at 2.9 kg (6.41 lb), took just forty-seven days and ate barely anything.[30] In the UK, we buy a slightly smaller chicken on average – at 2.2 kg (4.9 lb) – but only have to wait a brief thirty-five days to go from baby chick to chicken dinner.[31] Birds sold as 'extra-large' roasters often exceed knee-buckling weights of 4 kg (nearly 9 lb), enough to feed a family of six to eight people. To say that the lifespan of a broiler chicken is brief is an understatement; while a modern meat bird will enjoy just seven or so weeks of life, and is eaten as a super-sized toddler, a backyard hen can expect to live at least ten years, while the chicken's ancient ancestor, the red junglefowl, has been known to live for three decades.

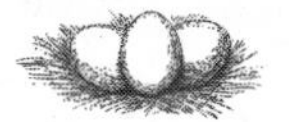

The high cost of research and development has meant that the poultry industry has become dominated by two major breeding companies - Cobb-Vantress and Aviagen. Anywhere in the world you buy, eat or cook a chicken, there's a strong likelihood that one of these two players was responsible for its genetic characteristics. Indeed, the process of creating and refining broiler breeds is so shrouded in secrecy that broiler birds are not known by their breed or cross-breed titles but by a brand name or trademark. Broilers have names like the Cobb500 or the Ross 308 and come from a long pedigree of birds.

These cross-breed chickens might make excellent broilers but they can't reproduce themselves with any reliability - when cross-breeds mate, the resulting chicks could have any number of traits of previous generations. To get a uniform broiler every time, therefore, poultry companies go through a complex process of generational crosses. Like a poultry royal family, it's all about the lineage. The breeding company will maintain a small nucleus of different pure-breed chickens to create the perfect crosses. From this small elite, it takes around four years of generational breeding - first great-grandparents, then grandparents, then parents, and finally the broiler offspring. Each generation grows exponentially in number, so a handful of elite chickens can end up providing the genetics for billions of broilers. These chicken 'family trees' can also be so complicated and trade-secret that the birds could never be reproduced outside high-level bio-secure genetic breeding facilities.

The creation of a broiler that can grow bigger, faster and on less food helped drive down the price of chicken for the consumer, transforming it from an occasional luxury to an everyday choice. Poultry was also the first agricultural sector to become an 'agri-business', one where a food company can own multiple stages in the process. In the early years of the broiler business, feed mills, hatcheries, farms and abattoirs or 'processors' were all separate businesses. Soon after the Chicken of Tomorrow contest, these separate stages began to be combined so that one large company or 'integrator' could run the show and, in the process, drive down costs. In the 1960s, the UK government began recording the average price of chicken. Taking into account inflation, an average broiler fifty years ago cost around £11 in today's money and yet supermarkets now sell medium whole chickens for less than the price of a takeaway latte.*

In a modern, integrated broiler firm, the process is owned from start to finish, apart from the initial breeding of the 'parent stock'. An intensively farmed broiler chicken will start its life as an egg laid in a breeding facility, where it will be transported to a separate hatchery. There, it will be artificially incubated for three weeks and, at a day old, the new chick will be moved again to a finishing unit or grow-out farm, where it will be raised with up to 30,000–50,000 other chicks in a climate-controlled

* At the time of writing, Aldi are selling a British medium whole chicken for £2.75. A Caffè Nero 'Regular Latte' is £3.15.

shed. After about seven weeks, the chicken will have reached its desired weight and will be sent to a processing factory. From there, the meat may go on to other factories, to be turned into different foods, while the by-products - feathers, blood, bone, skin, feet and other undesirables - are either incinerated or rendered into other, often surprising, things. More of that in the next chapter.

Poultry is now the most consumed meat across the globe, overtaking pork in the last few years. In 2021, the world noshed its way through over 130 million metric tonnes, compared to 112 million metric tonnes of pork and 70 million metric tonnes of beef. To feed that kind of demand, 66 billion chickens are slaughtered per year and, at any one point in the year, there is a live population of over 22 billion chickens,[32] or three birds for every person on the planet. The world population of chickens - a bird that started life as an isolated, exotic junglefowl - is also three times greater than *all the other wild bird species combined.*

We have also created a bird that is so far removed from its original morphology that it can no longer survive without human and technological intervention. The genetic selection that created birds that grow fat, fast, has also created a number of unpleasant health problems for the modern broiler. The chickens that are the most efficient at converting food into flesh also tend to have lower metabolic rates and low oxygen consumption, which makes them prone to heart failure and ascites or 'water belly', a disease that causes a build-up of fluid in the abdominal cavity. According to a report by the Royal Society, if left to live longer than a few weeks, the modern broiler is unlikely to survive into adulthood.[33] The genetic selection for rapid growth of leg and breast flesh has led to a relative decrease in the size of vital organs including the heart and lungs, which then affects a chicken's ability to function properly. The pneumatic body shape of the hefty modern broiler, with its huge breast, stumpy legs and low centre of gravity, also causes lameness and skeletal problems that affect the bird's ability to walk properly. One experiment, where broilers were allowed to live for nine weeks instead of five, resulted in seven times more birds dying before they reached their slaughter date.[34]

Modern broilers are both docile and voraciously greedy, a winning combination for an animal deliberately bred to be obese. Broilers that are kept for breeding, rather than eating, and are allowed to reach sexual maturity at about 15–18 weeks, must be kept on a severely restricted 'starvation' diet, otherwise they die prematurely or become too big to mate; this can leave birds, according to one research paper, 'chronically hungry, frustrated and stressed' and with high levels of male aggressiveness towards the females. In conventional UK broiler farming, which has better welfare standards than many other countries, the legal stocking density (the amount of birds you can have per square metre) is 33 kg, which equates to about fifteen average broiler birds. Given the floor space of a standard oven is about a quarter of a square metre, it seems the broiler chicken enjoys more room after it's dead than at any other point in its short life.

8

TRAILBLAZERS

'Farmaceuticals' and Chickens of the Future

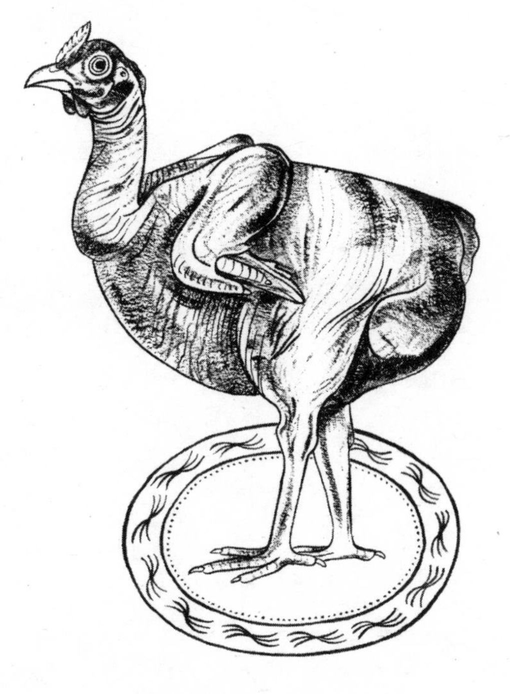

Naked or Featherless Chicken

The Hunterian Museum at the Royal College of Surgeons is not for the faint-hearted. Like a deliciously nightmarish sweet shop, the museum displays row upon row of glass jars, not of confectionery but of eighteenth-century anatomical specimens collected by the wild but brilliant father of scientific surgery, John Hunter. Among the thousands of macabre souvenirs, including a collection of hernias and the left half of mathematician Charles Babbage's brain, is the most curious specimen of all - a cockerel's head with a human tooth embedded into its comb.

Hunter, who was said to be the inspiration for Robert Louis Stevenson's *Dr Jekyll and Mr Hyde*, was fascinated by dissection and transplants. When he wasn't paying exorbitant prices for cadavers from illegal bodysnatchers, Hunter also conducted tests and autopsies on many of the animals he kept in a menagerie at his country home just outside London. There, both exotic and domestic creatures - everything from lions and leopards to sheep, dogs and pigs - were subjected, sadly often still kicking and squealing, to bizarre experiments. Chickens, it seems, proved a rich seam to mine. One of his first experiments with poultry involved cutting off a cockerel's spur and grafting it onto the bird's comb, where it miraculously adhered and continued to grow. Hunter had been inspired by an 'old and well-known' hoax among travelling shows, where unicorn-like chicken 'freaks' were created in the same way and paraded around for profit. A chicken's comb is rich in blood vessels, it turns out, and an excellent candidate for external grafting. Hunter, emboldened by his success with the 'cockerel's tooth', then attempted a different experiment, this time grafting a cockerel's testicle into its belly, before repeating the procedure but transplanting a cockerel's testicle into a hen. Incredibly, all his grafts took.

Hunter had also trained his sights on the world of dentistry, a profession that, in the eighteenth century, was not only disreputable but downright dangerous, despite the sums of money exchanging hands. Hunter was convinced he could find a way to transplant healthy human teeth into the mouths of well-off patients, but he first had to prove the principle. Hunter's *Treatise on the Natural History and Diseases of the Human Teeth*, published in 1778, described his attempts: 'I took a sound tooth from a person's head; then made a pretty deep wound with a lancet into the thick part of a cock's comb, and pressed the fang of the tooth into this wound, and fastened it with threads passed through other parts of the comb. The cock was killed some months after, and I injected the head with a very minute injection; the comb was taken off and put into a weak acid, and the tooth being softened by this means, I slit the comb and tooth into two halves, in the long direction of the tooth. I found the vessels of the tooth well injected, and also observed that the external surface of the tooth adhered everywhere to the comb by vessels, similar to the union of a tooth with the gum and sockets.' Hunter, however, did qualify his findings with a muted afterthought: 'I may here just remark, that this experiment is not generally attended with success. I succeeded but once out of a great number of trials.'

Hunter nonetheless went on to popularise tooth transplants. Fellow surgeon Thomas Bell later wrote in the notes to an 1839 reprint of Hunter's toothy treatise: 'The operation in question became a favourite one with him; and it appears to have been very frequently performed either by himself or under his directions.'[1] Poverty-stricken donors would be paid a small sum to have their healthy teeth pulled, without anaesthetic, which were then implanted into the gap-toothed gums of wealthy clients. While the transplanted teeth might stay in place for a few years, they rarely 'took' and were notorious for passing on disease. The procedure's repeated failure 'induced almost all subsequent practitioners to abandon its employment' and yet Hunter's chicken experiments proved an important stepping stone in the development and understanding of blood vessels and transplants. To this day, ceremonial gowns of the Royal College of Surgeons' Faculty of Dental Surgeons still feature an unusual logo in recognition of Hunter's eccentric but ground-breaking work – a beautifully embroidered, tooth-headed cockerel.

The chicken has, in fact, played a starring role in one of the world's greatest medical triumphs – the development of the vaccine. The aim of almost all vaccines is to trick the body's immune system into producing antibodies that fight pathogens, tiny organisms that cause disease and include viruses and bacteria. By introducing a weakened or deactivated version of the pathogen,* the body develops resistance to the disease without succumbing to the worst of its effects. Ancient civilisations are thought to have loosely understood this principle; historical records show that for hundreds of years, Chinese practitioners had used the dried scabs of smallpox victims by crushing them into dust and blowing them up the noses of healthy people, who would then only contract a mild form of the disease and become resistant to future infections. Another technique was to place powdered material or pus from a smallpox scab into a scratch in the skin, a practice reported in parts of Africa, the Middle East and Asia, news of which reached Europe via word of mouth and travellers' tales by the eighteenth century.

In England, country surgeon Edward Jenner noticed that milkmaids who picked up a mild disease called cowpox seemed to be immune to the much more deadly smallpox, which killed a third of its victims and often left survivors blind and with terrible scars. In 1796, he successfully infected an eight-year-old boy with cowpox introduced via a scratch and then, a few months later, exposed the same child to smallpox; he developed no serious effects. Jenner tested his hypothesis on a further twenty-three subjects, including his own baby son, Robert, and in 1801 published *On the Origin of the Vaccine Inoculation*. In it, he

* Modern vaccines actually work in a number of ways – some don't contain a live or weakened version of the virus and instead deliver a genetic code or 'blueprint' to the body's cells, giving instructions on how to build an immune response.

summarised his discoveries, maintaining that 'the annihilation of the smallpox, the most dreadful scourge of the human species, must be the final result of this practice'.[2] The medical establishment was initially slow to accept Jenner's findings, but by 1840 the British government had passed the Vaccination Act, which provided smallpox vaccinations free of charge. The subsequent 1853 Act made the vaccination compulsory for every child from birth.

Just two decades later, the French microbiologist Louis Pasteur was wrestling with the problem of chicken cholera, a disease that was ripping through the poultry population. Pasteur had been a great admirer of Jenner's work and believed that if a vaccine could be found for smallpox, a vaccine could be created for pretty much every other disease of man or beast. Pasteur was developing a vaccine for chickens, made from cultured cholera bacteria, but was finding that many of the birds died soon after the injection. The strength of the vaccine was too potent. It was then that he made an accidental breakthrough. Pasteur had instructed his assistant to inject a new batch of chickens before the assistant went away on holiday. The technician forgot, however, and only managed to inject the flock a month later, with the now weeks-old vaccine. After their shots, the chickens were only mildly ill and then recovered fully. Then, when given a subsequent fresh dose of cholera bacteria, none of the chickens became sick. Pasteur realised that the original cholera culture had been weakened over time, by exposure to air, and had become a safer and more effective vaccine as a result. And while the idea of using a diminished form of a disease to provide immunity wasn't new, it was Pasteur and his unwitting chickens who proved the theory of it in a laboratory setting. Over the next century, many more weakened or 'attenuated' vaccines were developed – including those for diphtheria, plague, tuberculosis, yellow fever, measles, mumps and rubella – saving millions of lives.

For the world to fully embrace vaccines, however, there had to be a reliable way to produce them in vast quantities and with decent consistency. Unlike bacteria, which can be grown in a laboratory in a suitable medium such as agar gel, viruses can't reproduce by themselves and need to infect living cells. After a virus infects a cell, it uses its own genetic instructions to make more copies of itself. Throughout the nineteenth century, scientists often had to grow viruses in live animal (and sometimes human) hosts,* which could then be harvested and made into safe vaccines. Smallpox was grown in the skin of calves, for instance, and rabies in rabbits. In the early years of

* In the early stages of smallpox vaccines, immunity would be spread from person to person; people infected with cowpox developed fluid-filled sores on their skin after a week or so, which would be lanced by a doctor and the 'serum' collected and given to an unvaccinated person via a scratch. The potency of the serum didn't last very long outside the body, however, so taking the vaccine abroad – to immunise colonists – was proving tricky. One shocking solution in the early years of the nineteenth century proved effective, however. Orphan children as young as three were pulled from institutions and made to act as 'vaccine chains' aboard ships; these live carriers of cowpox, who would be deliberately infected one after the other, at ten-day intervals, could keep the virus alive for months until the ship reached land.

the twentieth century, a handful of scientists were attempting to grow viruses in fragments of tissue, rather than in a live host, but kept coming up against two problems: one was that some viruses stubbornly refused to grow in an artificial setting and the second was that the tissue samples often became contaminated. Without a pure, sterile environment, bacteria and other pollutants were ruining results.

During the 1920s, a trio of scientists – the pathologist Ernest Goodpasture, and husband and wife team Eugene and Alice Woodruff – were working on viruses at Vanderbilt University in Nashville, Tennessee. Goodpasture had served as a Navy doctor during the First World War and had experienced first-hand the devastation wrought by the 1918 influenza epidemic, which infected around five hundred million people worldwide and killed about one in ten. Goodpasture was also convinced that, if vaccines were going to save people from another devastating pandemic, science was going to have to find a way to grow large amounts of viruses in cheap, sterile cultures, rather than relying on live animal hosts or expensive and contaminant-prone tissues.

While experimenting with the fowl pox virus, he hit upon the idea of using fertilised chicken eggs. As we already know, eggs are miraculous containers for an embryo, providing all the nutrients and conditions for growth; they're also the perfect, naturally sterile vessel to support a living organism. Goodpasture and the Woodruffs, after a number of failed attempts, devised a way to cut a small hole in the shell of an egg, inject a tiny amount of a virus into the membrane, cover the hole with glass and seal it with Vaseline.[3] The fertile eggs were then put into an incubator and within a few days the fowl pox virus had multiplied to such an extent that it could be harvested from each egg in a useful and, more importantly, untainted form. Buoyed by their success with fowl pox, Goodpasture and his associates turned their sights to other viruses. Along with herpes simplex, the team also managed to cultivate the smallpox virus, with each single egg miraculously producing enough viral material to vaccinate a thousand children. Other researchers soon followed suit, using the same technique, and within a few years mass vaccines had been developed for yellow fever, typhus and, eventually, influenza.[4]

At the beginning of the Second World War, the US government threw itself into funding and delivering the influenza vaccine; it was terrified of a repeat of the 1918 pandemic, which had hit military populations particularly hard. According to one estimate, influenza accounted for nearly 80 per cent of war casualties suffered by the US Army during the First World War[5] and was so deadly that American intelligence erroneously speculated it might have been deliberately unleashed by the Germans. Millions of refugees and military service members, huddled together in camps, ships, lorries and planes, were the perfect breeding ground for another disaster. Dr Thomas Francis Jr, chairman of the Influenza Commission during the Second World War, explained just how worried the American government were: 'The appalling pandemic of 1918 in the last months of the exhausting conflict of World War I, with massive mobilization of armies and upheaval of civilian populations, has irrevocably linked those two catastrophes. It demonstrated that virulent influenza may be more devastating to human life than war itself [...] the onset of another war inevitably recalled the spectre of 1918 and the possibility that [it] would again result in the epidemiologic conditions which would heighten the severity of influenza to a catastrophic level.'[6] Thanks to the initial discovery by Goodpasture, and subsequent scientists who refined and developed the flu shot, in 1945 the US Army vaccinated eight million troop members and the much-feared post-war pandemic was avoided. The following year, the egg-based vaccine was made available to the general public and by the early 1950s deaths from flu in the US had halved.[7]

Although science eventually came to realise that new flu vaccines would have to be formulated every year to protect against the different strains, Goodpasture's work changed the course of medical history. Chicken eggs have been used to create seasonal and pandemic influenza vaccines ever since, saving countless lives in the process. In the US alone, around half a billion eggs are employed every year in the production of vaccines.[8] In an interesting twist to the tale, governments are now looking into non-egg alternatives; a pressing problem is that the supply of egg-based vaccines is now vulnerable to increasing outbreaks of avian influenza or 'bird flu'. Eggs used in the manufacture of vaccines come from a handful of secret farms, where poultry is raised specifically to lay high-quality eggs. Over the past few decades, the world's poultry population has exploded – between 1990 and 2007, chicken consumption around the globe more than doubled[9] and, as we have seen, there are more than twenty billion chickens alive at any one time.

According to the World Organisation for Animal Health, the upsurge in waves of avian flu in recent years has been linked to changes in agricultural practices, such as high density of flocks and genetic homogeneity, the intensification of the poultry sector and the globalisation of trade in live poultry.[10] And, as a recent report to Congress by the US Government Accountability Office conceded, it only takes one case of bird flu to wipe out vast numbers of chickens, regardless of whether they're diseased or not: 'Because of the environment in which commercial birds are raised, if one bird becomes infected with a notifiable avian influenza, hundreds of thousands of birds can be exposed and will need to be depopulated.'[11] Chickens raised for laying vaccine eggs are kept in bio-secure locations but nowhere is completely immune from the threat of bird flu. If the highly sensitive supply chain for vaccine eggs is compromised by an outbreak, it could mean a national shortage of influenza shots. There's a depressing circularity to the fact that scientists are now racing to roll out a flu shot for chickens to safeguard the future of the human influenza vaccine.

While some manufacturers are looking into alternatives to the egg-based vaccine, science isn't finished with the egg quite yet. Lots of human diseases are caused by the body not naturally producing enough of something or by manufacturing it incorrectly. This can be a protein, for example, or an enzyme. Some of these illnesses can be controlled with drugs but these are often cripplingly expensive or difficult to produce on a commercial scale. Scientists recently discovered that genetically modified chickens can be used to lay eggs that contain these missing substances. Genetically modified or 'transgenic' chickens have had their DNA altered in some way while they were still embryos. The chicken then grows into an adult who lays eggs containing certain therapeutic agents; researchers at Roslin Institute in Scotland, for instance, have successfully modified chickens to produce large amounts of a specific protein in their egg whites, a process that is significantly cheaper than making synthetic proteins in a laboratory.[12] So far, their research has focused on two different proteins - one has antiviral and anti-cancer properties, and the other is being developed to help damaged tissue repair itself. Similar studies elsewhere have created hens' eggs that produce monoclonal antibodies, molecules that mimic the human systems to fight specific pathogens, and enzymes that could potentially treat lysosomal acid lipase deficiency, a life-threatening condition that causes fat to accumulate in the body.

One of the most unusual uses of genetic modification of hens' eggs, however, has nothing to do with the world of medicine. Researchers at Charles Sturt University in Australia have been attempting to develop an egg that glows green. While this may seem, on the face of it, like indulgent meddling, the research is attempting to solve one of the most serious ethical and economic problems with current world egg production.

As we've already learned, prior to the early twentieth century and the development of separate broiler and egg-laying breeds, in any given flock the cockerels would be kept for meat (and breeding) and the females used for eggs. Once the chicken industry developed different hybrid birds for abundant eggs and fast-growing meat, cockerels at an egg-laying farm suddenly became redundant.

To have an egg industry, you need a never-ending supply of new hens. The hatcheries who supply egg producers with chicks, however, have a pressing problem about what to do with the unwanted males. Until recently, there was no commercially viable way for sexing embryos *in ovo* - *i.e.* inside the egg; producers had to wait until all their chicks hatched and then keep the girls but dispatch the boys. Few people realise that the current method of killing these one-day-old chicks is either to gas them *en masse* or, more horrifyingly, shred them alive. Macerating or 'live-blending' chicks, as it's known in the industry, is a widespread practice that involves sending chicks along a conveyor belt into a high-speed grinder, to end up as fertilisers, animal feed and pet food. About seven billon male chicks are thought to be culled every year.[13]

Animal welfare groups and enlightened producers and retailers have recently turned their attention to creating 'no-cull' eggs. To do this, producers have to accurately sex and destroy male eggs before the nerve endings of an embryo form, just a few days into their incubation, rather than wait until they hatch; a number of techniques for *in ovo* sexing are starting to be used, including near-infrared light beams (a technique called spectroscopy) or testing each embryo with a tiny amount of fluid called a bio-marker, which changes colour depending on the sex of the chick. The researchers at Charles Sturt University in Australia, however, have been trying a different approach. They have isolated a gene from a jellyfish that fluoresces green under ultraviolet light. By inserting this gene into one of the sex chromosomes of a chicken so that only female embryos have it, it's hoped that simply scanning eggs with a laser that detects fluorescence will enable producers to automatically detect the green embryo of a female and discard any eggs that fail to glow.[14]

The commercial poultry industry produces a whole host of unwanted by-products, not just the unfortunate male chick. Feathers from the broiler industry have proved particularly problematic. Of the 10,000 tonnes of waste feathers that are produced globally *every day*,[15] only a tiny fraction are used in their intact form, either as a low-quality stuffing (goose and duck down are softer and more insulating than chicken feathers) and in craft or decorative items. The vast majority of the billions of chicken feathers plucked every year are either incinerated, which adds to global carbon emissions, turned into low-grade animal feed* or ends up as landfill, where it can cause soil and water pollution.

* Feathers have even been turned into human food. A recent Central Saint Martins student, Sorawut Kittibanthorn, found a way to turn chicken feathers into a cheap source of edible protein. In his experiments, Kittibanthorn created feather-based pasta and protein bar biscuits. He also tested different binding agents to create a product with the right colour and consistency to mimic red meat.

One startling solution has been the development of the featherless or 'naked chicken'; although yet to be a widespread commercial success, the development of a broiler bird that comes practically 'oven ready' has become a reality. From an economic point of view, the idea makes sense – no feathers, no waste, less cost. Naked broilers also seem to show a higher meat yield, because they're not using energy growing feathers, and have been touted as a solution to the problem of intensive poultry farming in hot countries, where birds often suffer and die from heat stress. The featherless chicken is, in fact, not a genetically modified bird but one of two naturally occurring but rare mutations. Naked Neck chickens have been a recognised breed for decades – Naked Neck bantams were first recorded at the 1898 German National Show – and have a gene that reduces their plumage by around 20–40 per cent, especially around their necks and bottoms. A different mutation produces a bird called the 'scaleless chicken', a sorry creature that lacks any body feathers as well as foot scales and spurs and is to all intents and purposes completely starkers.

Research at the Hebrew University of Jerusalem has been looking into the practical advantages of breeding Naked Neck broilers in hot climates, as well as introducing the scaleless gene (which is mostly found in egg layers) into fast-growing broiler breeds to create a featherless meat chicken that can tolerate high ambient temperatures.[16] And while the economics make sense, it's hard not to see the bird as a misguided solution to a horrible problem. Detractors claim that naked chickens, without the protective barrier of their plumage, are more prone to parasites, insect bites, sores and sunburn. Removing the feathers also makes mating difficult for both sexes - in the flurry of copulation, a cockerel will struggle to hang on to a bald female with his claws and beak, and ends up injuring his partner in the process. A cockerel's mating ritual - which involves plenty of glorious wing flapping and plumping of feathers - is also rendered useless if the bird is naked. The fact that we are making efforts to alter the chicken's natural form, rather than changing the damaging and unpleasant way we farm the bird, speaks volumes about how skewed our approach to food production has become.

Feathers are made up of 90 per cent keratin, the key protein that gives nails, horns, hooves, claws, hair and scales their inherent toughness. Keratin fibres are also hollow, which gives feathers both their lightness and their insulating qualities. Feathers also don't combust as readily as some synthetic products. Engineers and scientists from different countries have been coming up with novel ways to utilise these natural properties and create useful and often biodegradable products from the mountains of feather waste produced every year, including building insulation, bioplastics, packaging and flame-retardant coatings. Researchers in America found that feather fibre (essentially powdered feathers) could be added to plastic used in car parts, such as dashboards, to increase its strength while reducing its overall weight. Feather fibre is finding its way into paper, cosmetics, nappies, shoe soles, circuit boards and plant pots,[17] and is even being trialled as a potential material for aircraft and space travel.

In fact, the world of aerospace isn't just interested in chicken feathers. Many of us will have heard of Laika, the Soviet space dog, or Ham, the first chimpanzee astronaut, just two of the many animals involuntarily dragged into man's space ambitions. But spare a thought for Kentucky, a chicken who spent his early life orbiting Earth over eighty times. The story begins in 1979, when Russians sent Japanese quail eggs into space to see what the effect of zero gravity would be on developing embryos. If birds could hatch in space, pondered the scientists, they could potentially be a future food source for astronauts. The experiment wasn't an initial success but by 1990 the Russians had managed to hatch a quail chick on the *Mir* space station, the first vertebrate ever to have been born in space. More quails followed but it seemed that none could cope with the peculiarity of weightlessness – researchers noticed the female quails stopped ovulating and neither sex showed any interest in mating without the grounding effects of gravity. The idea of breeding poultry in space for food was not looking promising.

Meanwhile, back at NASA, American scientists were turning their attention to chicken eggs. A high-school student, John Vellinger, had pitched the idea of taking fertilised eggs in an incubator into space, to see what effect zero gravity had on their development. On Earth, the force of gravity on a yolk pushes it towards to the edge of the shell – backyard hens and industrial egg producers both must turn their eggs to stop the growing chick sticking to the shell. For sponsorship for the project, NASA and Vellinger approached the fast-food chain KFC, who were keen to put their name to the high-profile experiment. The first attempt ended in disaster - the space-bound eggs were put aboard the ill-fated Space Shuttle *Challenger,* which exploded just a few minutes after take-off. Three years later, NASA and Vellinger tried again - this time putting thirty-two fertilised eggs on the Space Shuttle *Discovery*, an endeavour nicknamed 'Chix in Space'. Half of the eggs had been fertilised nine days before launch, the other half just two days prior to lift-off, to test whether the maturity of the embryo had any effect on its ability to survive a period of weightlessness. The eggs went on a journey of a lifetime, orbiting two million miles around the Earth over the space of five days, before touching back down on dry land. Less than a week after *Discovery* landed, the first space-chick hatched at KFC's headquarters in Louisville. Named, unsurprisingly, Kentucky, the cockerel and seven other chicks from the older egg group survived, while none of the eggs fertilised just two days before take-off developed properly. The experiment raised more questions than it answered - gravity clearly played a key role in the healthy growth of a new embryo but some of the 'older' embryos had coped with a short period of weightlessness. Space-station chicken farming was clearly some way off.

And yet that's not quite the end of the story. While a live chicken is yet to soar to the clouds, in 2017 KFC once again set its sights on the skies in a mission of breath-taking folly. In collaboration with World View Enterprises, the stratospheric exploration company, KFC launched its 'Zinger 1 Space Mission' with the lofty ambition to 'take KFC's new spicy, crispy Zinger chicken sandwich to new heights'. The breaded chicken breast fillet, lettuce and mayo sandwich was carried high into the stratosphere on a space-proof KFC bucket, pulled by a helium-filled balloon. KFC planned for the Zinger 1 craft, floating silently at the edge of space, to take a selfie using a robotic arm, drop a KFC coupon to Earth, wave a KFC flag and, of course, tweet. What seems like a mad-cap plan was actually designed to promote high-altitude balloon technology, which could eventually spend months in the stratosphere at a time, providing communication and edge-of-space-tourism. In the end, the chicken sandwich had its maiden voyage cut short seventeen hours into its planned four-day trip, thanks to a small balloon leak. The finances involved in attempting to make a dead chicken fly to the edge of space can only be guessed at, but space product placement isn't actually a new idea. Back in 1996, Pepsi paid for Russian cosmonauts to pose with a four-feet-tall replica of a Pepsi can while they went on a spacewalk outside the former *Mir* space station, while in 2001 Pizza Hut delivered a six-inch salami pizza to the International Space Station.

Sticking with the skies, chickens have also been front and centre in the scientific study of tornados. Throughout the nineteenth century, eminent American meteorologists were attempting to understand just exactly what happened when a twister struck. The United States and Canada have more tornados than any other countries in the world, and studying their movements and mysterious patterns has always been a challenge. Along with some of the more obvious questions – such as what conditions lead up to a tornado and what happens in the eye of a column – a rather more unusual phenomenon was piquing one scientist's interest. Numerous newspaper reports from the time reported that chickens were often left completely naked when a tornado passed over. The mathematician and weather fanatic Professor Elias Loomis wanted to find out why. Describing one twister in 1837, Loomis noted there was a tremendous roar and 'Several of the fowls were picked almost clean of their feathers, as if carefully done by hand.'[18] Loomis included the event in his list of the eleven things most likely to characterise a tornado.

The defeathering of live chickens raised some interesting questions for Loomis – if it was a certain wind speed that was plucking the feathers from the birds, could it be reproduced and measured, giving us an indication of the wind speed inside a tornado? To test the idea, Loomis loaded a small cannon with a dead chicken, feathers intact, and blasted it skywards. Loomis recorded the results with not a little excitement: 'The feathers rose twenty or thirty feet, and were scattered by the wind. On examination, they were found to be pulled out clean, the skin seldom adhering to them. The body was torn into small fragments, only a part of which could be found. The velocity was 341 miles per hour. A fowl, then, forced through the air with this velocity is torn entirely to pieces; with a less velocity, it is probable most of the feathers might be pulled out without mutilating the body.' From this rather crude experiment, Loomis determined that most tornadoes must spin slower than 341 mph but fast enough to cleanly pluck the feathers out; he speculated that if a live bird was fired from a cannon at just 100 mph, that would probably do the job. Thankfully, Loomis never tested his hypothesis, but it was still believed well into the mid-twentieth century that winds stripped chickens of their plumage. A few other scientists had postulated different theories over the decades; one was that a tornado's electric charge might be stripping the feathers, another that a vacuum created in the column would somehow cause the shafts of the feathers to explode, but no one really came up with a convincing alternative.

Then, in 1975, Bernard Vonnegut, scientist and older brother of the novelist Kurt, looked into the veracity of Loomis's experiment by teaching himself about how a bird's plumage works. He discovered that a chicken's follicles hold on to its feathers more or less tightly depending on the bird's health, nervous state and whether it's moulting. This meant it was impossible to use the plucking of a chicken as an accurate indicator of tornado wind speed. Vonnegut then suggested a different explanation for the mystery of the twister-plucked chicken. When poultry are highly stressed, they can perform a 'flight moult', where their feathers fall out more easily than usual. This biological response has evolved to help chickens survive being grabbed by a predator, leaving the attacker with a mouthful of loose feathers and allowing the bird to escape. Chickens, in the eye of a storm, were perhaps loosening their feathers as part of an innate stress-response mechanism, allowing the tornado to do the rest.[19]

While Vonnegut was posthumously awarded the Ig Nobel Prize, a satirical honour for scientific achievements that make people both laugh and think, one famous scientific experiment involving a chicken didn't end so happily. Francis Bacon was born in 1561, became a Member of Parliament in 1584 and Lord Chancellor in 1618. But his real passion lay in science. In 1626, Bacon was journeying in a coach towards Highgate in London with the king's Scottish physician, Dr Witherborne. It was an unseasonably cold spring and snow was still on the ground. Bacon, who was fascinated with the 'conservation and induration* of bodies',[20] suddenly hit upon the idea that flesh might be preserved in snow, as it was in salt. Both men excitedly jumped from the coach and, according to the account of the episode in John Aubrey's *Brief Lives*, 'went into a poore woman's howse at the bottome of Highgate Hill, and bought a hen, and made the woman exenterate it, and then stuffed the bodie with snow, and my lord did help to doe it himselfe'.[21] According to a letter written by Bacon to his friend the Earl of Arundel, 'As for the experiment itself, it succeeded excellently well: but in the journey (between London and Highgate) I was taken with such a fit of casting, as I knew not whether it were the stone, or some surfeit, or cold, or indeed a touch of them all three. But when I came to your lordship's house, I was not able to go back, and therefore was forced to take up my lodging here, where your house-keeper is very careful and diligent about me.' Bacon didn't realise just how poorly he was, and his chill quickly

* Induration – the hardening of soft tissue or organs.

turned into something much more serious. 'I know how unfit it is for me to write to your lordship with any other hand than my own; but by my troth my fingers are so disjointed with this fit of sickness that I cannot steadily hold a pen',[22] he bravely scribbled in what turned out to be the last letter he ever wrote. In just a few days, Bacon was dead.

Although his bizarre demise inspired many a witty pun, about frozen chicken lasting longer than bacon, this curious episode in history has a rather extraordinary postscript. Since Bacon's death, there have been a handful of 'sightings' of a ghostly chicken haunting Pond Square in Highgate. One newspaper report, from 1957, described the 'shivering chicken' as a regular visitor that residents had often seen, including Mr and Mrs John Greenhill, who insisted: 'It was a big whitish bird and it used to perch on the lower boughs of the tree opposite our house [...] Many members of my family have seen it on moonlit nights' before it 'disappeared through a brick wall'.[23] Surely a poultry-geist if ever there was one.

We're at an interesting juncture in the long and bumpy story of the chicken. A recent report produced by the United Nations and the Organisation for Economic Co-operation and Development (OECD) predicted that global meat consumption will grow by 12 per cent over the next decade, with cheap chicken meat accounting for half of the increase. It projected that, by 2030, the average amount of meat consumed per person, globally, every year will be over 35 kg. Chicken will constitute over 40 per cent of this yearly meat consumption, twice as much as beef and eight times as much as lamb or mutton.

Two key factors are driving this rise in production and consumption: one is that modern chicken broiler breeds, out of all the animals farmed for meat, are particularly good at converting what they eat into flesh. Producers talk a lot about Feed Conversion Ratios (FCRs), and chickens, compared to say cows, give a farmer the same amount of meat for much less animal feed. Chickens also grow more quickly than they ever have done in history – this short production cycle, from egg to dinner plate in just a few weeks, means that producers can endlessly tweak and refine chickens' genetics, diet and living environments to be ever more efficient. As we saw with the 'naked chicken' research project, many breeding companies are now looking at creating birds that can cope with hostile environments or lower-quality feeds – conditions that can be more common in less economically developed nations.

The second factor is that the pattern of meat consumption is expected to change over the next ten years. Population growth will be one of the main pressures on demand – by 2030, the Earth's inhabitants are predicted to be about a tenth greater in number than at present. Certain parts of the world are growing faster than others, however. That is why meat consumption is projected to grow by 30 per cent in Africa, 18 per cent in Asia, 12 per cent in Latin America and 9 per cent in North America but only 0.4 per cent in Europe. But lots of factors affect how much meat a country eats – from income (meat is traditionally expensive) to religious beliefs, welfare concerns to health trends. Currently, per capita meat consumption is low in developing countries but near saturation levels in more developed nations, where we have been munching large amounts of meat for decades. Changing consumer preferences, such as a growing interest in vegetarian or vegan diets, and meat substitutes, are also influencing demand in developing countries.[24]

Meat consumption is still, however, at its highest in high-income countries, with wealthy nations eating staggering amounts compared to the poorest. The average European eats nearly 80 kg a year, the average American 110 kg and the average Australian 116 kg. Compare this with some of the poorest African countries – such as Mozambique and Niger – who eat as little as 10 kg per year per person and it's easy to see how inequitable the situation is. Countries who have experienced rapid economic growth over the past decades have also seen an explosion in the consumption of meat, especially chicken. China now eats fifteen times more meat than it did in 1961.[25] Rates in Brazil have quadrupled. India is one of the only countries to buck the trend – thanks to a population who are largely vegetarian or lacto-vegetarian, its per capita meat consumption is tiny at 4.5 kg, a rise of only 1 kg in two decades.[26]

The ethics of the global chicken industry are a muddle. High-income nations have enjoyed the nutritional benefits of cheap chicken and eggs for decades and it seems hypocritical to tell low-income nations what they can and can't grow and eat. But increases in chicken production come at a cost. According to the UN, livestock is responsible for nearly 15 per cent of all man-made greenhouse emissions – equivalent to all the world's cars, planes, boats and trains – and a third of all the crops we grow are fed directly to animals. The picture is complex. If a supermarket can sell a whole chicken for less than the price of a latte, whoever is being paid to raise those birds is receiving considerably less. The margins for many farmers are tight. Cost efficiency has been one of the prime movers in industrial chicken production for decades. The goal seems an impossible one – to grow increasingly larger chickens, more quickly, and with a decreasing input in terms of feed, energy and other costs. The sophistication of animal genetic development and technological solutions seems to know no bounds and, on the face of it, poultry that is raised intensively is more efficient in terms of land use, food resources and carbon emissions than more traditional, free-range systems or other forms of livestock. Grass-fed, slowly maturing cows release the equivalent of 16 kg of carbon dioxide for every kilogram of meat produced, while factory-farmed chickens are responsible for a paltry 4.4 kg of CO_2 per kilo.[27]

But 'efficiency' isn't the only measure of environmental sustainability. According to the Livestock, Environment and People Project (LEAP), a collaboration between a number of different organisations including the University of Oxford and the International Food Policy Research Institute, a more nuanced approach must take into consideration all the other issues created by intensive systems. An increase in the demand for all types of meat, including chicken, has had a profound effect on deforestation around the world, which is itself a major source of greenhouse gases and a dire threat to biodiversity. In the past, much of this tree-felling was for beef, but more recently the demand has come from soya beans to feed poultry. The Sustainable Food Trust estimates that the UK imports around 3 million tonnes of soya annually, the majority of which goes to chickens. And while its production has been clearly linked to deforestation in South America, over half of the soya used to feed poultry in the UK is not certified deforestation-free.[28]

Intensive chicken production can also be a major source of pollution, particularly the creation of mountains of manure and feathers. This is often used as fertiliser, which, although packed with nutrients and a valuable agricultural product, comes with its own problems. The excessive or indiscriminate application of chicken waste on farmland can end up causing more harm than good, polluting watercourses and introducing antibiotics, trace elements, pesticides and pathogens into the environment and, potentially, the food chain.

The widespread use of antibiotics in intensive chicken production has also been linked to the growing problem of antibiotic resistance, where some strains of bacteria become immune to treatment. Antibiotic-resistant bacterial infections are now a leading cause of death worldwide, according to *The Lancet*, killing about 3,500 people every day, more than HIV or malaria combined.[29] The global poultry industry has also been criticised for poor working conditions, especially in processing factories and abattoirs, while the appearance of new, intensive production sites in traditional farming communities is often blamed for noise and air pollution, increases in HGV traffic, damage to local rivers and the industrialisation of a landscape that also relies on tourism. It's worth saying that many chicken producers and farmers are working hard to make improvements - in the UK, antibiotic use in livestock has fallen 50 per cent over the last five years and the EU recently banned the administration of antibiotics to healthy animals, including poultry - but there's still a long way to go.

Perhaps most important, however, is the increasing disconnect that seems to exist between our affection for creatures, great and small, and the treatment of intensively farmed animals, especially chickens, as a commodity to be exploited rather than sentient, living beings. Our relationship with the chicken is so contradictory: we coo over baby chicks but order fast-food legs and wings by the bucket; during lockdown, we consoled ourselves by panic-buying backyard hens, only for many of us to dump them when life returned to normal; and the chicken is simultaneously a symbol of traditional farm life, scratching around in a bucolic existence, and the pin-up for factory-farmed cruelty. It seems ignorance is bliss.

Thanks to high-profile campaigns about the horrors of battery cages, nearly half of all the eggs produced in the UK are now free-range, but few consumers know where their chicken meat comes from; 94 per cent of all chicken consumed in the UK is from intensively reared birds. But things are changing. Thanks to the efforts of groups such as Compassion in World Farming, consumers are now finding higher welfare alternatives on supermarket shelves, which provide chickens with more space, natural light and a stimulating environment. Labelling is key - terms such as 'farm fresh' or 'corn fed' are meaningless when it comes to raising happy chickens – but other terms such as 'free-range', 'certified organic' or 'enhanced welfare' are useful.*

But does any of this matter? Do chickens really care how they're kept? And more to the point, should we? The commoditisation and treatment of chickens has, ever since we first domesticated them, mostly gone unquestioned. From cock-fighting to capons, ancient sacrifice to battery eggs, it seems our capacity for cruelty knows no bounds. Even modern-day attitudes towards farm animals are both complex and contradictory. There's a whole raft of literature and study dedicated to the ethics of meat eating, much of which is brilliantly tackled in *The Meat Paradox* by Rob Percival (2002), but at the heart of the debate is the fundamental conflict between our enjoyment of and desire to eat meat and our moral opposition to animal cruelty and suffering.

On the face of it, people have two choices – one is to decide that they don't like animals all that much anyway, so don't care how they're treated, and the other is to go vegetarian or vegan. In reality, most of us do neither and instead opt for a third strategy

* It's a sliding scale; in the UK, for example, the Soil Association label ensures that chickens have space to move around, that they are able to perch, peck and play, have natural daylight, live most of their lives outdoors and that the farmer has raised slower-growing breeds. The 'RSPCA Assured Free-Range' label indicates the same high welfare standards. The 'RSPCA Assured' label ticks all the same boxes apart from the free-range status, meaning the birds live indoors, as does the 'Red Tractor Enhanced' sticker. At the lower end of the scale, the normal Red Tractor label – which is the common food safety standard required by most UK retailers – ensures that the producer meets basic animal welfare standards currently in line with the law. Today, this means high stocking densities of fast-growing breeds who live a life completely indoors, although Red Tractor allow 10 per cent more space than is required by European legislation and are phasing in the requirement for natural light.

– to disengage with any discomforting feelings using a number of tricks or habits. This is called 'cognitive dissonance'. We buy meat that bears no resemblance to the animal it once was. We choose not to ask about the conditions our meat animals are raised in. We even mentally create different categories for pets and farm animals, each group with its own perceived intelligence and capacity for suffering. In reality, these categories are spurious – pigs have been shown to be as intelligent as dogs, hens can learn certain tasks more quickly than toddlers. We use the insult 'chicken-brained' without understanding the breadth of their cognition or behavioural abilities.

The chicken is actually a remarkable creature. Studies have shown that, contrary to what many people believe, there's plenty going on behind those beady eyes. Experiments demonstrate that chickens possess a whole host of important visual, mental and spatial capacities, including 'object permanence', the ability to understand that something exists even when out of sight. In other words, if you show a chicken an object and then take it away, it still knows it exists; baby chicks have this skill almost as soon as they've hatched, while in human babies it takes about three months to develop.[30]

Newly hatched chicks have also been shown to be capable of simple maths, solving addition and subtraction problems with numbers from zero to five.[31] Incredibly, chicks also seem to count upwards moving left to right – smaller numbers on the left, larger numbers to the right; this left-right mental 'number line' is the same way humans picture numbers. Chickens have also demonstrated a surprising level of self-control – in experiments where the longer the hens waited, the larger their food reward, chickens not only showed the ability to recognise that the outcomes would be different depending on their patience but were able to employ willpower and hold out for the biggest rewards. These amazing birds have also been shown to be capable of judging specific time intervals and anticipating future events.[32]

Perhaps most extraordinary, however, is their ability to communicate and express emotions. Chickens have been shown to have a repertoire of at least two dozen distinct vocalisations, as well as numerous different visual displays. Referential communication is something we imagine only humans to have – this is where specific meaning is attached to a signal or vocalisation. For years, we thought animals only expressed very simple communications that contained low levels of information, such as fear or aggression. Referential communication is interesting because, in the same way that humans use words for objects, chickens use different calls for different threats – rather than have just one warning call for 'predator', they have a range of vocalisations for terrestrial and aerial predators. The calls are also subtly different within each of these categories, depending on the animal attacker. Cockerels doing a 'tidbitting' call – where they alert the hens to food – also vary their vocalisations based on the quality of the meal they've found. And, as we saw at the end of Chapter 4, chickens can even be Machiavellian with their communication, with cockerels making false food calls to lure females and alarm calls that put male rivals in mortal danger.

Chickens are also capable of social learning – picking up

skills by observation rather than trial and error. In one test, young hens who watched trained hens perform a task were able to perform that task correctly more often than young hens who'd watched similarly inexperienced hens attempt the same task. Chicks have been shown to be able to discriminate between different human faces; one extraordinary experiment even seemed to show that chickens prefer pictures of 'beautiful' humans with symmetrical faces, rather than human faces with more asymmetrical features; humans show the same preference, keying in on symmetry as one of the subconscious measures of health and, therefore, mate attractiveness.[33]

Studies have shown our feathered friends to be capable of fear, anxiety, empathy and emotional distress on behalf of their chicks.[34] Anyone who has ever looked after a flock of chickens will know each bird has its own personality or temperament. Within these complex groups, it's easy to identify the individuals who are bold or more reserved, inquisitive or nervous, vigilant or wonderfully blithe. Some are aggressive or resist human contact, others enjoy being fussed over. And while chickens might not be the most demonstrative of pets, there's an increasing understanding of the beneficial relationship that can exist between human and fowl. 'Therapy chickens' – birds brought into care homes, hospitals and other therapeutic settings – are now a widely accepted tool to help people with a complex range of needs from autism to loneliness, depression to post-traumatic stress disorder.

So where does all this leave our relationship with our fowl friends? The fact that chickens have been shown to have unexpected cognitive abilities and social relationships doesn't necessarily mean we shouldn't eat their eggs or farm them for meat. That's a personal choice and one that comes down to how you view our role in the food chain. It should, however, make us think about how we treat chickens in terms of their physical and mental well-being, and their ability to engage in natural behaviour such as perching, dust-bathing and nesting. As one report in *Scientific American* concluded after reviewing studies on poultry intelligence: 'Chickens are smart, and they understand their world, which raises troubling questions about how they are treated on factory farms.'[35]

Chickens are the animal that epitomises the careless and, increasingly, dangerous relationship we have cultivated with the rest of the living world. Chickens and eggs have proved to be an incredibly valuable food source for humans. At one end of the scale, these magnificent creatures represent the ideal animal for self-sufficient, slow living – intelligent, biddable, generous and full of character. Conversely, as a global society we have also become particularly good at treating chickens as objects in a system of food production that disregards their feelings and capacity for pain or misery. Chickens have been 'de-animalised'; in many people's minds they are not even viewed as 'birds' who need to express their natural and social nature, but are rather commodities to be grown, like plants, and 'harvested'. We've played a terrible trick on the chicken, and few give it the respect or compassion it deserves. It's fowl play indeed.

ACKNOWLEDGEMENTS

As always, a million thanks go to the plucky team at Head of Zeus who helped bring *Fowl Play* to life and to those who will make it fly: Matt Bray, Kathryn Colwell, Clémence Jacquinet, Claire Kennedy, Aphra Le Levier-Bennett, Ed Pickford, Matilda Singer and, most of all, Richard Milbank.
You're all very good eggs indeed.

Silkie (Chinese Silk Chicken)

ENDNOTES

Chapter 1

1. Lee, M. S. Y et al., 'Sustained miniaturization and anatomical innovation in the dinosaurian ancestors of birds', *Science*, 345:6196 (1 Aug 2014), pp. 562–6.
2. Field, D. J. et al., 'Early Evolution of Modern Birds Structured by Global Forest Collapse at the End-Cretaceous Mass Extinction', *Current Biology*, 28:11 (4 June 2018), pp. 1825–31.e2.
3. Field, D. J., Benito, J., Chen, A. et al., 'Late Cretaceous neornithine from Europe illuminates the origins of crown birds', *Nature*, 579 (2020), pp. 397–401. https://doi.org/10.1038/s41586-020-2096-0
4. Larson, D. W. et al., 'Dental Disparity and Ecological Stability in Bird-like Dinosaurs prior to the End-Cretaceous Mass Extinction', *Current Biology*, 26:10 (2016), pp. 1325–33. https://doi.org/10.1016/j.cub.2016.03.039
5. Callaway, E., 'T. rex kinship with chickens confirmed', *New Scientist* (24 April 2008). www.newscientist.com/article/dn13772-t-rex-kinship-with-chickens-confirmed/#ixzz6w4RWKlLk
6. Romanov, M. N., Farré, M. et al., 'Reconstruction of gross avian genome structure, organization and evolution suggests that the chicken lineage most closely resembles the dinosaur avian ancestor', *BMC Genomics*, 15:1060 (2014). https://doi.org/10.1186/1471-2164-15-1060
7. Grossi, B. et al., 'Walking Like Dinosaurs: Chickens with Artificial Tails Provide Clues about Non-Avian Theropod Locomotion', *PLoS ONE* 9(2):e88458 (2014). https://doi.org/10.1371 journal.pone.0088458
8. Bhullar, B. A. S. et al., 'A molecular mechanism for the origin of a key evolutionary innovation, the bird beak and palate, revealed

by an integrative approach to major transitions in vertebrate history', *Evolution*, 69:7 (July 2015), pp. 1665–77. https://doi.org/10.1111/evo.12684

9. www.scientificamerican.com/article/mutant-chicken-grows-alli/
10. Coghlan, A., 'Female Ducks Fight Back Against "Raping" Males', *New Scientist* (1 May 2007). www.newscientist.com/article/dn11764-female-ducks-fight-back-against-raping-males/?ignored=irrelevant
11. Hopkin, K., 'Why Did the Chicken Lose Its Penis?', *Scientific American* (6 June 2003). www.scientificamerican.com/podcast/episode/why-did-the-chicken-lose-its-penis-13-06-06/
12. McKinnell, J., 'Cock-a-doodle-don't', Macleans (20 March 2006). https://archive.macleans.ca/article/2006/3/20/cock-a-doodle-dont
13. Comte de Buffon, G. L. L., *The Natural History of Birds: From the French of the Count de Buffon; Illustrated with Engravings, and a Preface, Notes, and Additions, by the Translator,* Vol. II (Cambridge: Cambridge University Press, 2010), p. 54.
14. Ibid., p. 111.
15. Blyth. E., 'XXXIX – A Few Critical Remarks on M. Carl J. Sundevall's Paper on the Birds of Calcutta', in *The annals and magazine of natural history, zoology, botany and geology: incorporating the journal of botany*, Vol. 20 (London: Taylor and Francis, 1847), p. 388.
16. Blyth, E., 'Letter on the poultry in India', Gardener's Chronicle & Agric. Gaz., 39:619 (1851), as quoted in Grouw, H. and Dekkers, W., 'Temminck's *Gallus giganteus*; a gigantic obstacle to Darwin's theory of domesticated fowl origin?', *Bulletin of the British Ornithologists' Club*, 140:3 (21 Sep 2020), pp. 321–34.
17. Blyth, E.. Letter to C. R. Darwin, 30 September or 7 October 1855. Darwin Correspondence Project, letter no. 1761, as quoted in Grouw, H. and Dekkers, W., 'Temminck's Gallus giganteus; a gigantic obstacle to Darwin's theory of domesticated fowl origin?', Bulletin of the British Ornithologists' Club, 140:3 (21 Sep 2020), pp. 321–34.
18. Darwin, C. R., 'Fowls', in *The variation of animals and plants under domestication* (London: John Murray, 1st ed., 2nd issue. Vol. 1, 1868), p. 225. www.darwin-online.org.uk

19. Ibid., pp. 236–7.
20. Fumihito, A. et al., 'One subspecies of the red junglefowl (Gallus gallus gallus) suffices as the matriarchic ancestor of all domestic breeds', *Proceedings of the National Academy of Sciences*, 91:26 (Dec 1994), pp. 12505–9. https://doi.org/10.1073/pnas.91.26.12505
21. Eriksson, J., Larson, G., Gunnarsson, U. et al., 'Identification of the yellow skin gene reveals a hybrid origin of the domestic chicken', *PLoS Genet,* 4(2):e1000010 (29 Feb 2008). https://doi.org/10.1371/journal.pgen.1000010
22. www.scientificamerican.com/article/earliest-chickens-were-actually-pheasants/
23. Wang, M. S., Thakur, M., Peng, M. S. et al., '863 genomes reveal the origin and domestication of chicken', *Cell Research*, 30 (2020), pp. 693–701. https://doi.org/10.1038/s41422-020-0349-y
24. Clutton-Brock, J., *The Walking Larder: Patterns of Domestication, Pastoralism, and Predation* (London: Routledge, 2014), p. 11.
25. Serpel, J. A., 'Pet-Keeping in Non-Western Societies: Some Popular Misconceptions', *Anthrozoös*, I:3 (1989), pp. 170–1.
26. Collias, N. E. and Saichuae, P., 'Ecology of the Red Jungle Fowl in Thailand and Malaya with Reference to the Origin of Domestication', *Natural History Bulletin of the Siam Society*, 22 (1967), pp. 189–209.
27. Ibid.
28. Agnvall, B., 'Early domestication? Phenotypic alterations of Red Junglefowl selected for divergent fear of humans', Linköping Studies in Science and Technology, Dissertation No. 1790, IFM Biology, Department of Physics, Chemistry and Biology, Linköping University, Sweden (2016). www.diva-portal.org/smash/get/diva2:1036456/FULLTEXT01.pdf

Chapter 2

1. Perry-Gal, L. et al., 'Earliest economic exploitation of chicken outside East Asia: Evidence from the Hellenistic Southern Levant', Proceedings of the National Academy of Sciences, 112:32 (July 2015), pp. 9849–54. https://www.jstor.org/stable/26464787
2. Carter, H., 'An Ostracon Depicting a Red Jungle-Fowl. (The Earliest Known Drawing of the Domestic Cock)', Journal of

Egyptian Archaeology, 9:1/2 (1 Apr 1923), pp. 1–4. https://doi.org/10.2307/3853489

3. Ryley Scott, G., *History of Cockfighting* (Hindhead: Saiga Publishing, 1983), p. 91.
4. Bostock, J. (ed.), Plin. Nat. 10.25, in *Pliny the Elder, The Natural History* (London: Taylor and Francis, 1855). http://data.perseus.org/citations/urn:cts:latinLit:phi0978.phi001.perseus-eng1:10.25
5. Pegge, Rev. S., 'A Memoir on Cockfighting', *Archaeologica,* Vol. III (The Society of Antiquaries of London, 1786), p. 141.
6. Csapo, E., 'The Cultural Poetics of the Greek Cockfight', The Australian Archaeological Institute at Athens, Bulletin 4 (2006/07 [2008]) 20–37, 2006), p. 25.
7. Macurdy, G. H., 'The Derivation and Significance of the Greek Word For "Cock"', *Classical Philology*, 13:3 (1918), p. 311.
8. Csapo, E., 'The Cultural Poetics of the Greek Cockfight', The Australian Archaeological Institute at Athens, Bulletin 4 (2006), p. 25.
9. Geoffrey Arnott, W., *Birds in the Ancient World from A to Z* (London: Routledge, 2007).
10. Bostock, Plin. Nat. 30.49, in *Pliny the Elder, The Natural History*.
11. Ibid., Plin. Nat. 10.25.
12. Watson, L. C., *Magic in Ancient Greece and Rome* (London: Bloomsbury Academic, 2019).
13. http://data.perseus.org/citations/urn:cts:greekLit:tlg0525.tlg001.perseus-eng1:2.34
14. Cassius Dio, 'Epitome of Book LXXVII', in *Roman History,* published in Vol. IX of the Loeb Classical Library edition (1927), ch. 15. https://penelope.uchicago.edu/Thayer/E/Roman/Texts/Cassius_Dio/77*.html
15. As quoted in McMahon, J., Paralysin Cave: Impotence, Perception and Text in the Satyrica of Petronius (Leiden: Brill, 2018–1998?), p. 132.
16. Bostock, Plin. Nat. 37.54 in *Pliny the Elder, The Natural History*.
17. Pliny, as quoted in Scott, *The History of Cockfighting*, p. 88.
18. Radin, M., 'The Lex Pompeia and the Poena Cullei', *Journal of Roman Studies,* 10 (1920), pp. 119–30. https://doi.org/10.2307/295798
19. Polybius, 'Smoking Out the Enemy', in *Histories*, Book 21.28.

www.perseus.tufts.edu/hopper/text?doc=Perseus%3Atext%3A1999.01.0234%3Abook%3D21%3Achapter%3D28

20. Caesar, J., *Commentarii de Bello Gallico* (Caesar's Gallic War), W. A. McDevitte (trans.), 1st ed. (New York: Harper & Brothers, 1869), Book 5, Chapter 12.
21. Doherty, S. P. et al., 'Estimating the age of domestic fowl (*Gallus gallus domesticus* L. 1758) cockerels through spur development', *International Journal of Osteoarchaeology*, Volume 31, Issue 5, (Sep/Oct 2021), pp. 1–12. https://doi.org/10.1002/oa.2988
22. Best, J., Feider, M. and Pitt, J., 'Introducing Chickens – arrival, uptake and use in prehistoric Britain', *PAST: The Newsletter of the Prehistoric Society*, 84 (Autumn 2016).
23. Grossi, F., 'Through Celts and Romans: Technology and Symbolism of Bronze Enameled Roosters' (Università degli Studi di Milano 2017, J. Paul Getty Trust). www.getty.edu/publications/artistryinbronze/statuettes/19-grossi
24. Livarda, A. et al., *The Bioarchaeology of Ritual and Religion* (Barnsley: Oxbow Books, 2017).
25. Columella, L. J. M., *De Re Rustica*, Fe. S. Forster, M.B.E., M.A.(Oxon.), F.S.A. (trans.) (Cambridge, Mass: Harvard University Press, 1954), Book VIII, Chapter II.
26. Wood-Gush, D. M. G., 'A History of the Domestic Chicken from Antiquity to the 19th Century', *Poultry Science*, 38:2 (March 1959), pp. 321–6. https://doi.org/10.3382/ps.0380321
27. Kron, G., 'Poultry Farming', in Lindsay Campbell, G. (ed.), *The Oxford Handbook of Animals in Classical Thought and Life* (Oxford: OUP, 2014), pp. 119–21.
28. Columella, L. J. M., *De Re Rustica*, Book VIII, Chapter V.
29. Ibid.
30. Doucleff, M., 'How The Sweet Potato Crossed The Pacific Way Before The Europeans Did', NPR (23 January 2013). www.npr.org/sections/thesalt/2013/01/22/169980441/how-the-sweet-potato-crossed-the-pacific-before-columbus?t=1625579601284
31. Young, E., 'Polynesians beat Columbus to the Americas', *New Scientist*, (4 June 2007). www.newscientist.com/article/dn11987-polynesians-beat-columbus-to-the-americas/#ixzz6zv5sm75H
32. Taonui, R., 'Te Haerenga Waka: Polynesian Origins, Migrations,

and Navigation', thesis for MA in Maori Studies, University of Auckland (1994).

33. Pigafetta, A., 'The Philippine Islands, 1493–1898. Explorations by early navigators, descriptions of the islands and their peoples, their history and records of the catholic missions, as related in contemporaneous books and manuscripts, showing the political, economic, commercial and religious conditions of those islands from their earliest relations with European nations to the close of the nineteenth century', Vol XXXIII, 1519–1522, Blair, E. H. (ed.) (Cleveland, OH: Arthur H. Clark Company, 1906). https://www.gutenberg.org/files/42884/42884-h/42884-h.htm
34. Marsden, as quoted by Rennie, J. in *The Domestic Habits of Birds* (Charles Knight 1883), p. 103.
35. Wood, Jamie R. et al., 'Origin and timing of New Zealand's earliest domestic chickens: Polynesian commensals or European introductions?', *Royal Society Open Science*, 3:8 (2016).
36. Ibid, p. 40.
37. Ibid.
38. Thomson, G. M., *The Naturalisation of Animals and Plants in New Zealand* (Cambridge: Cambridge University Press, 2011), p. 15.

Chapter 3

1. Hoggard, B., *Magical House Protection: The Archaeology of Counter-Witchcraft* (New York: Berghahn Books, 2019).
2. *The most strange and admirable discoverie of the three Witches of Warboys arraigned, convicted and executed at the last Assises at Huntington, for the bewitching of the five daughters of Robert Throckmorton Esquire, and divers other persons, with sundrie Divelish and grievous torments: And also for the bewitching to death of the Lady Crumwell, the like hath not bene heard of in this age* (London: Printed for Thomas Man and John Winnington, and are to be solde in Paternoster Row, at the signe of the Talbot, 1593)., p 55. https://archive.org/details/WarboysWitchesTrialAndExecution/page/n55/mode/2up
3. Guiley, R. E., *The Encyclopedia of Witches, Witchcraft and Wicca* (New York: Infobase Publishing, 2008), p. 44.

4. *The Dublin University Magazine*, Vol. 10 (William Curry, Jun., and Company, 1837), p. 451.
5. Ibid.
6. Browne, Sir T., 'A Miscellany of Mistaken Beliefs', in *Pseudoxia Epidemica*, Book V, ch. 22. http://penelope.uchicago.edu/pseudodoxia/pseudo522.html
7. Scot, Reginald, *The Discoverie of Witchcraft* (1584), ch. 4. www.gutenberg.org/files/60766/60766-h/60766-h.htm
8. Bishop, Prof. T. et al., *The Shakespearean International Yearbook: Volume 15: Special Section, Shakespeare and the Human* (Farnham: Ashgate Publishing, Ltd, 2015), p. 172.
9. Mills Alden, H. (ed.), *Harper's Magazine*, Vol. 69 (New York: Harper's Magazine Company, 1884), p. 100.
10. *Edinburgh Magazine* (Feb 1818), p. 117.
11. Grimm, W. and J., '*Die Wichtelmänner: Drittes Märchen, Kinder- und Hausmärchen*', no. 39/III, Ashliman, D. L. (trans.). https://sites.pitt.edu/~dash/gerchange.html
12. Jacobs, J., 'The Brewery of Eggshells', in *Celtic Fairy Tales* (London Pan Macmillan, 2011).
13. GideonBohak,G.,'TraditionsofMagicinLateAntiquity–Protective Magic – Babylonian Demon Bowls' (The Michigan Society of Fellows and Department of Classical Studies, December 1995). https://deepblue.lib.umich.edu/bitstream/handle/2027.42/108169/def2.html
14. As quoted in Brand, J., *Observations on Popular Antiquities: Chiefly Illustrating the Origin of Our Vulgar Customs, Ceremonies, and Superstitions* (London: Charles Knight and Company, 1841), p. 99.
15. Ibid., p. 98.
16. Newall, V., 'Easter Eggs', *The Journal of American Folklore*, 80:315 (1967), pp. 3–32. https://doi.org/10.2307/538415
17. https://blogs.bl.uk/digitisedmanuscripts/2017/04/a-hunt-for-medieval-easter-eggs.html
18. Brand, J., *Observations on popular antiquities: including the whole of Mr. Bourne's Antiquitates vulgares*, revised by Sir Harry Ellis, Vol. 1 (1849), p. 176.
19. www.hebdenbridgehistory.org.uk/folklore/pace-egg2.html
20. https://blogs.loc.gov/folklife/2016/03/easter-bunny/

21. https://blog.english-heritage.org.uk/the-history-of-the-egg-hunt/
22. Ibid.
23. https://blogs.bl.uk/asian-and-african/2019/02/jewish-love-potions-a-users-guide.html
24. Radford, E. and M. A., *Encyclopaedia of Superstitions* (London: Rider and Company, 1947), p. 112.
25. Keats, J., 'The Eve of St Agnes', in *The Poetical Works of John Keats* (1884). www.bartleby.com/126/39.html
26. Kightly, C., *The Perpetual Almanac of Folklore* (London: Thames & Hudson, 1987). 'March 2: St Chad's Day: eggs now become plentiful'.
27. Klein, J. A., 'Cock Ale: "A Homely Aphrodisiac"'. https://recipes.hypotheses.org/3018
28. www.smithsonianmag.com/smart-news/meet-pro-temperance-women-who-crusaded-against-coffee-180965039/
29. Comte de Buffon, G. L. L., *The Natural History of Birds: From the French of the Count de Buffon; Illustrated with Engravings, and a Preface, Notes, and Additions, by the Translator*, Vol. II (Cambridge: Cambridge University Press, 2010).
30. As quoted in Hartley, D., *Food in England* (London: Piatkus, 2009), p. 181.
31. E. P. Evans as quoted in Walter, E. V., 'Nature on Trial: The Case of the Rooster That Laid an Egg', *Comparative Civilizations Review*, Vol. 10:10, Article 7 (1985). https://scholarsarchive.byu.edu/ccr/vol10/iss10/7
32. Lepp, A. J., 'The Rooster's Egg: Maternal Metaphors and Medieval Men' (University of Toronto, 2010). https://tspace.library.utoronto.ca/bitstream/1807/26509/6/Lepp_Amanda_J_201011_PhD_thesis.pdf
33. Zeitlyn, D., 'Do Mambila Cockerels Lay Eggs? Reflections on Knowledge and Belief', *JASO* 22/1 (1991), pp. 59–64. www.anthro.ox.ac.uk/sites/default/files/anthro/documents/media/jaso22_1_1991_59_64.pdf
34. *The Satyricon of Petronius Arbiter: Complete and unexpurgated translation by W. C. Firebaugh, in which are incorporated the forgeries of Nodot and Marchena, and the readings introduced into the text by De Salas*. https://www.gutenberg.org/files/5225/5225-h/5225-h.htm

35. Saint John Chrysostom, *Commentary on the Epistle to the Galatians, and Homilies on the Epistle to the Ephesians, of S. John Chrysostom* (Parker, 1879), p. 242.
36. Dhalla, M. N., *Zoroastrian civilization: from the earliest times to the downfall of the last Zoroastrian empire, 651 A.D.* (New York: OUP, 1922), p. 185.
37. Ibid., p. 66.
38. Quoted in Hulme, F. E., *The history, principles and practice of symbolism in Christian art* (London: S. Sonnenschein; New York: Macmillan, 1892), p.191.
39. Simpson, J., *Folklore of Sussex* (Cheltenham: The History Press, 2009).
40. Rhŷs, J., *Celtic Folklore: Welsh and Manx*, Vol 1 (Oxford: OUP, 1901), p. 85.
41. Shakespeare, W., *Hamlet*, Act 1, Scene 1.
42. Dryden, J., *The Poetical Works of John Dryden: With a Life, Critical Dissertation, and Explanatory Notes By the Rev. George Gilfillan*, Vol. II (Edinburgh, 1855). www.online-literature.com/dryden/poetical-works-vol2/
43. 'An Enquiry into the Original Meaning of Cock-Throwing on Shrove-Tuesday', *Gentleman's Magazine* (London, 1736–50), 7 (Jan 1737), pp. 6–8. https://web.archive.org/web/20110424093330/http://www.animalrightshistory.org/animal-rights-timeline/animal-rights-g/gentlemans-magazine/1737-01-enquiry.htm
44. Ibid.
45. Griffin, E., '*Popular Recreation and the Significance of Space*', *British Academy Review*, 9. www.thebritishacademy.ac.uk/documents/619/11-griffin.pdf(17 Mar 2017).
46. Henderson, W., *Notes On The Folk Lore Of The Northern Counties Of England And The Borders, With an appendix on Household stories by S. Baring-Gould* (London: Longmans, Green and Co. 1866), p.78.
47. *The Youth's Cornucopia* (London: Hamilton, Adams and Company, 1832), p. 251.

Chapter 4

1. Morris, W., 'The Story of Rhodope', in *The Earthly Paradise* (London and New York: Routledge, 2014).

2. Stoker, B., *Dracula* (1897), ch. 12. *'Letter, Mina Harker to Lucy Westenra 17 September'* (https://www.gutenberg.org/files/345/345-h/345-h.htm)
3. Major, C., *Dictionary of Afro-American Slang* (New York: International Publisher's Co., Inc., 1970).
4. Grose, F., *Dictionary In The Vulgar Tongue: A Dictionary Of Buckish Slang, University Wit, And Pickpocket Eloquence* (1819 edition). www.gutenberg.org/cache/epub/5402/pg5402.html
5. Butler, S., *The Genuine Remains in Verse and Prose of Mr. Samuel Butler, Author of Hudibras: Published from the Original Manuscripts, Formerly in the Possession of W. Longueville, Esq;* (London: J. and R. Tonson, 1759), p. 104.
6. Dryden, J., 'Pastoral III', in *The Works of John Dryden, by Walter Scott, Esq.*,Vol. XIII (Edinburgh: James Ballantyne and Co., 1808). https://www.gutenberg.org/files/47383/47383-h/47383-h.htm
7. www.bbc.co.uk/arts/robertburns/works/the_henpecked_husband/
8. 'Hen *n.1*', *Dictionary of the Scots Language* (2004). www.dsl.ac.uk/entry/snd/hen_n1
9. Skelton, J., 'Colyn Cloute', in *The Poetical Works of John Skelton*,Vol 1 (London: T. Rodd, 1843).
10. Howitt, W., *Land, Labour, and Gold: Two Years in Victoria: With Visits to Sydney and Van Diemen's Land*, Vol. 2 (Cambridge: Cambridge Library Collection – History of Oceania, 2011), p. 140.
11. 1880 *Bulletin* (Sydney), 17 July, as quoted in https://slll.cass.anu.edu.au/centres/andc/meanings-origins/c
12. https://johnsonsdictionaryonline.com/views/search.php?term=chuck
13. 1728 Ramsay Poems II. 226. www.dsl.ac.uk/entry/snd/chuck_n1
14. Brontë, E., *Wuthering Heights*, ch. 34. www.gutenberg.org/files/768/768-h/768-h.htm
15. Doherty, S. P. et al., 'Estimating the age of domestic fowl cockerels through spur development', International Journal of Osteoarchaeology (2021), p. 9. http://clok.uclan.ac.uk/37551/8/37551%200a.2988%20%281%29.pdf
16. Bardsley, C. W., *Curiosities of Puritan Nomenclature* (London: Chatto and Windus, 1880).
17. Gard, A. J., 'The Rise of the Coquette in Seventeenth- and

Eighteenth-Century French Theater', dissertation for PhD in the Department of Romance Studies (University of North Carolina, 2017), p. 14. https://cdr.lib.unc.edu/downloads/cz30pt43t?locale=en

18. Davidson, J. P., *Planet Word* (London: Penguin Books Ltd, 2011).
19. www.poetryfoundation.org/poems/43702/the-four-ages-of-man-56d22282befab
20. "Poor Richard, 1739," *Founders Online,* National Archives, https://founders.archives.gov/documents/Franklin/01-02-02-0046. [Original source: *The Papers of Benjamin Franklin,* vol. 2, *January 1, 1735, through December 31, 1744*, ed. Leonard W. Labaree. New Haven: Yale University Press, 1961, pp. 217–28.]
21. Warren, A., 'The Orphic Sage: Bronson Alcott', *American Literature*, 3:1 (1931), pp. 3–13. https://doi.org/10.2307/2919951
22. Wagner, F., 'Eighty-Six Letters (1814–1882) of A. Bronson Alcott (Part One)', *Studies in the American Renaissance* (1979), pp. 239–308. www.jstor.org/stable/30227466
23. Flint, J., *Letters from America, containing observations on the climate and agriculture of the western states, the manners of the people, the prospects of emigrants, &c* (Edinburgh: W. & C. Tait, 1822), pp. 263–4.
24. McArthur, T. B., *Concise Oxford Companion to the English Language* (Oxford: OUP, 2005).
25. Tusser, T., *Five Hundred Points of Good Husbandry* (London: Richard Tottill, 1573). 'The good motherly Nurserye'. http://name.umdl.umich.edu/A14064.0001.001
26. As quoted in www.phrases.org.uk/meanings/cock-a-hoop.html
27. 'A Philological Fragment', in *The Knickerbocker*, 32 (1848), p. 388.
28. de Luna, J., *Dictionary of Inn-Sign Names in Medieval and Renaissance England* (portions of this previously published as 'Designators in Inn-Sign Names in Medieval and Renaissance England' in the 2015 KWHSS Proceedings) (2017). https://heraldry.sca.org/kwhss/2017/inn%20signs%20dictionary%20JdL.pdf
29. Aronson, J., 'Chickenpox', *BMJ* (Clinical research ed.), 2000;321:682 (2000). https://doi.org/10.1136/bmj.321.7262.682
30. Knowles, J., et al., *A Critical Pronouncing Dictionary of the English Language, Incorporating the Labours of Sheridan and Walker, So Far as Their Examples are in Accordance with the True Principles*

of Orthoepy and Established Usage: and Comprising Above Fifty Thousand Additional Words: Also a Key to the Pronunciation of Classical and Scripture Proper Names (Henry Bohn, 1850).

31. Grose, *Dictionary In The Vulgar Tongue: A Dictionary Of Buckish Slang, University Wit, And Pickpocket Eloquence* (1819 edition). www.gutenberg.org/cache/epub/5402/pg5402.html
32. Browne, Sir Thomas, *Pseudodoxia Epidemica III. Chapter XXVIIIi* (1646; 6th ed., 1672), pp. 206–9. https://penelope.uchicago.edu/pseudodoxia/pseudo328.html
33. Webster, N., *Dissertations on The English Language: With Notes, Historical And Critical. To Which Is Added, By Way of Appendix, An Essay on A Reformed Mode of Spelling, With Dr. Franklin's Arguments On That Subject* (Boston: Thomas & Company, 1789), p. 123. www.gutenberg.org/files/45738/45738-h/45738-h.htm
34. *The American Housewife: Containing the Most Valuable and Original Receipts in All the Various Branches of Cooker : and Written in a Minute and Methodical Manner: Together with a Collection of Miscellaneous Receipts, and Directions Relative to Housewifery: Also the Whole Art of Carving* by Experienced Lady (Dayton and Saxton, 1841). 'To Boil Eggs', p. 33. https://www.gutenberg.org/files/28452/28452-h/28452-h.htm
35. *The Hull Packet*, 14 March 1851, p. 4.
36. As quoted in https://wordhistories.net/2021/03/14/headless-chicken/
37. *Alexandria Gazette* (Alexandria, Virginia), 6 Jul 1868, p. 2.
38. Kim, T. and Zuk, M., 'The effect of age and previous experience on social rank in female red junglefowl, *Gallus gallus spadiceus*, *Animal Behaviour*, 60:2 (2000), pp. 239–44. https://doi.org/10.1006/anbe.2000.1469
39. Kokolakis, A. et al., 'Aerial alarm calling by male fowl (*Gallus gallus*) reveals subtle new mechanisms of risk management', *Animal Behaviour*, 79:6 (2010), pp. 1373–80.

Chapter 5

1. As quoted in Jacques, D., *The Ferme Ornée* (Staffordshire Gardens Trust, 1999). www.researchgate.net/publication/340704680_The_Ferme_Ornee

2. Pückler-Muskau, Hermann, as quoted by The Gardens Trust. https://thegardenstrust.blog/2018/03/31/eggscellent-eggsamples-of-eggscentricity/
3. https://britishlistedbuildings.co.uk/101053640-fowl-house-approximately-10-metres-to-west-of-vauxhall-farmhouse-tong
4. www.discoveringtong.org/discoveringtong.htm
5. *Illustrated London News*, Saturday, 23 December 1843, p. 9.
6. Darwin, C. R., *The variation of animals and plants under domestication*, 1st ed., 1st issue, Vol. 2 (London: John Murray, 1868), p. 45.
7. *The Field*, Saturday, 26 February 1853, p. 13.
8. Lewer, S. H. (ed.), *Wright's Book of Poultry* (London and New York: Cassell, 1912), p. 240.
9. Ibid., p. 241.
10. *The Times*, 13 January 1853, p. 4.
11. *The Field*, Saturday, 26 February 1853, p. 13.
12. *Punch*, Introduction to Vol. XXIV (Jan–Jun 1853).
13. *Punch*, Vol. XXV (Jul–Dec 1853).
14. Burnham, G. P., *The History of The Hen Fever: A Humorous Record* (Boston, MA: James French and Co., 1855), p. 112. https://www.gutenberg.org/files/40872/40872-h/40872-h.htm
15. Tegetmeier, W., *Poultry for the Table and Market Versus Fancy Fowls* (London, 1898), p. 8.
16. Ferguson, G., *Illustrated Series of Rare and Prize Poultry* (London, 1854). www.biodiversitylibrary.org/bibliography/152121
17. Scott, G. R., *The History of Cockfighting* (Hindhead: Saiga Publishing, 1983), p. 160.
18. Comte de Buffon, G. L. L., *The Natural History of Birds: From the French of the Count de Buffon; Illustrated with Engravings, and a Preface, Notes, and Additions, by the Translator*, Vol. II (Cambridge: Cambridge University Press, 2010).
19. https://heritagepoultry.org/blog/hamburg
20. As quoted in Burnham, *The History of The Hen Fever*, p. 133.
21. Ibid., p. 134.
22. Ibid., p. 21.
23. Ibid., p. 309.
24. *The Field*, Saturday, 26 February 1853, p. 13.
25. *The Times*, 13 January 1853, p. 4.

Chapter 6

1. www.science.org/news/2018/08/these-eggs-are-putting-new-spin-how-shells-get-their-shape
2. www.princeton.edu/news/2017/06/22/hatching-new-hypothesis-about-egg-shape-diversity
3. Aristotle, *Historia Animalium*, 6:2. http://classics.mit.edu/Aristotle/history_anim.6.vi.html
4. Diodorus Siculus, *Library of History*, Book I, 69–98. https://penelope.uchicago.edu/thayer/e/roman/texts/diodorus_siculus/1d*.html#note5
5. Réaumur, R-A. Ferchault de, *The art of hatching and bringing up domestic fowls, by means of artificial heat / Being an abstract of Monsieur de Reäumur's curious work upon that subject, communicated to the Royal Society. Translated from the French*, Trembley, F. R. S. (ed.) (London: C. Davis, 1750), pp. 2–3.
6. *The Travels of Sir John Mandeville: The version of the Cotton Manuscript in modern spelling. With three narratives, in illustration of it, from Hakluyt's 'Navigations, Voyages & Discoveries'* (London: Macmillan and Co. Limited; New York: The Macmillan Company 1900), p. 33. www.gutenberg.org/files/782/782-h/782-h.htm
7. Excerpt from *Utopia*, Book 2. www.luminarium.org/renlit/utopiafarming.htm
8. Landauer, W., 'The Hatchability of Chicken Eggs as Influenced by Environment and Heredity' (Storrs Agricultural Experiment Station, 1961), p. 28. https://opencommons.uconn.edu/saes/1
9. Ashley, S., 'The Vulgar Mechanic and His Magical Oven: A Renaissance alchemist pioneers feedback control', *Nautilus* (2014). https://nautil.us/the-vulgar-mechanic-and-his-magical-oven-rp-2783/
10. Réaumur, as quoted in Terrall, M., *Catching Nature in the Act: Réaumur and the Practice of Natural History in the Eighteenth Century* (Chicago: University of Chicago Press, 2014).
11. Gaultier, Dr J. F., as quoted in Landauer, 'The Hatchability of Chicken Eggs as Influenced by Environment and Heredity', p. 39.
12. Lewer, S. H., *Wright's Book of Poultry* (London, New York, Toronto and Melbourne: Cassell and Company, Limited, 1912), p. 88.
13. Geyelin, G. K., *Poultry Breeding in a Commercial Point of View, as carried out by the National Poultry Company (Limited) Bromley,*

Kent, 6th ed. (New York, 1881; 1st ed. 1867).
14. Landauer, 'The Hatchability of Chicken Eggs as Influenced by Environment and Heredity', p. 28.
15. *The Cyphers Incubator Catalogue and Guide to Poultry Culture* (New York: Wayland, 1899).
16. Meall, L. A., *Moubray's Treatise on Domestic and Ornamental Poultry*, New edition (London: Arthur Hill, Virtue and Co., 1854), p. 47.
17. Columella, L. J. M., *De Re Rustica*, Forster, E. S., M.B.E., M.A. (Oxon.), F.S.A. (trans.) (Cambridge, MA: Harvard University Press, 1954), Book VIII, Chapter II.
18. Loog, L. et al., 'Inferring Allele Frequency Trajectories from Ancient DNA Indicates That Selection on a Chicken Gene Coincided with Changes in Medieval Husbandry Practices', *Molecular Biology and Evolution*, 34:8 (Aug 2017), pp. 1981–90. https://doi.org/10.1093/molbev/msx142
19. Slavin, P., 'Chicken Husbandry in Late-Medieval Eastern England: *c.*1250–1400', *Anthropozoologica*, 44:2 (2009), pp. 35–56. https://doi.org/10.5252/AZ2009N2A2
20. Warren, E., *200 Eggs a Year Per Hen: How to Get Them* (Hampton, NH: Warren, 1904), p. 3.
21. Dryden quoted in John, F. J. D., 'OSU's world-record-breaking chicken sparked a fowl feud with newspaper'. www.mckenzieriverreflectionsnewspaper.com/story/2021/09/09/history/osus-world-record-breaking-chicken-sparked-a-fowl-feud-with-newspaper/3927.html
22. W. C. Conner quoted in John, 'OSU's world-record-breaking chicken sparked a fowl feud with newspaper'.
23. Ibid
24. Davenport, C. B., *Inheritance in Poultry* (CreateSpace Independent Publishing Platform, 2018), p. 5.
25. Robinson, L. in *Modern poultry husbandry* (1948), as quoted in Sayer, K., '"His footmarks on her shoulders": the place of women within poultry keeping in the British countryside, *c.*1880 to *c.*1980', *Agricultural History Review*, 61: II (2013), pp. 301–29. https://bahs.org.uk/AGHR/ARTICLES/61_2_5_Sayer.pdf
26. Ott, W. H., 'Criteria of Vitamin D Deficiency in Mature Chickens', MSc thesis submitted to the Oregon State Agricultural College,

June 1936, p. 7. https://ir.library.oregonstate.edu/downloads/8c97ks777

27. www.britishpathe.com/video/hen-spectacles (1951)
28. Touchette, J., 'Egges or Eyren?: The "boundless chase"', Saint Louis University Library Special Collections (October 2016). https://pius.slu.edu/special-collections/?p=4717

Chapter 7

1. *De Re Coquinaria of Apicius* (Walter M. Hill, 1936). https://penelope.uchicago.edu/Thayer/E/Roman/Texts/Apicius/6*.html#IX
2. www.medievalists.net/2020/11/13th-century-scandinavian-cookbook/
3. www.thepoultrysite.com/news/2014/08/how-much-chicken-do-we-eat-in-the-uk
4. Moubray, B., *A Practical Treatise on breeding, rearing and fattening all kinds of Domestic Poultry, Pheasants, Pigeons and Rabbits, including an account of the Egyptian method of hatching eggs by artificial heat, Fourth edition: with additions, on the management of swine, on milch cows and on bees* (London, 1830), p. 12. https://wellcomecollection.org/works/db336wdv
5. Slavin, P., 'Chicken Husbandry in Late-Medieval Eastern England: *c.*1250–1400', *Anthropozoologica*, 44:2 (2009), pp. 35–56. https://doi.org/10.5252/AZ2009N2A2
6. Ibid.
7. Pegge, S., *The Forme of Cury, A Roll of Ancient English Cookery Compiled, about A.D. 1390* (London, 1780). https://www.gutenberg.org/cache/epub/8102/pg8102.html
8. Tirel, G., *Le Viandier de Taillevent*. www.cornishpasties.org.uk/origins/pasty-3-historical/pasty-origins3-LeViandier.htm
9. Gessner, as quoted in Cvjetkovic, V. et al., 'Capons: a history of "horned" egg incubators and chick carers', *Wiener tierärztliche Monatsschrift*, 104 (2017), pp. 363–75.
10. Slavin, P., 'Chicken Husbandry in Late-Medieval Eastern England: *c.*1250–1400', *Anthropozoologica*, 44:2 (2009), pp. 35–56. https://doi.org/10.5252/AZ2009N2A2
11. Ibid.
12. Bostock, J., 'Who First Invented the Art of Cramming Poultry:

Why the First Censors Forbade This Practice', *Pliny The Elder, The Natural History* (London: Taylor and Francis, 1855), Book 10, Chapter 71.

13. May, R., *The Accomplisht Cook, or The Art and Mystery of Cookery* (London: R. Wood, 1665), p. 456.
14. *Harleian MS.279*, 'A Forme of Cury Douce', MS.55 (1450).
15. https://www.theatlantic.com/technology/archive/2013/11/perhaps-the-strangest-photo-youll-ever-see-and-how-its-related-to-turduckens/281852/
16. Shakespeare, W., *As You Like It*, Act 2, Scene 7 (Folger Shakespeare Library). https://shakespeare.folger.edu/downloads/pdf/as-you-like-it_PDF_FolgerShakespeare.pdf
17. Onions, C. T., 'Capon', in *A Shakespeare Glossary* (Oxford: Clarendon Press, 1911). www.perseus.tufts.edu/hopper/text?doc=Perseus:text:1999.03.0068:entry=capon
18. Hartley, D., *Food in England* (London: Piatkus, 2009), pp. 186–8.
19. Dunne, J. et al., 'Reconciling organic residue analysis, faunal, archaeobotanical and historical records: Diet and the medieval peasant at West Cotton, Raunds, Northamptonshire', *Journal of Archaeological Science*, 107 (2019), pp. 58–70. https://doi.org/10.1016/j.jas.2019.04.004
20. Boyce, C. and Fitzpatrick, J., *A History of Food in Literature from the Fourteenth Century to the Present* (London: Taylor and Francis, 2017), p. 24.
21. Moubray, *A Practical Treatise on breeding, rearing and fattening all kinds of Domestic Poultry, Pheasants, Pigeons and Rabbits*, p. 2.
22. As quoted by Hardouin de Péréfixe de Beaumont in *Histoire du Roy Henry le Grand* (Paris: Edme Martin, 1662).
23. Horowitz, S., *Putting Meat on the American Table* (Baltimore, MD: Johns Hopkins University Press, 2006), p. 103.
24. Tanner, G. C., *Consular Reports: Commerce, Manufactures, Etc. Volume 8* (United States Bureau of Foreign Commerce, 1882), pp. 39–41.
25. US Bureau of the Census, *Historical Statistics of the United States, Colonial Times to 1957* (Washington, DC, 1960), pp. 122–3.
26. Platt, F. L. as quoted in Squier, S. M., *Poultry Science, Chicken Culture. A Partial Alphabet.* (Brunswick, NJ: Rutgers University Press, 2011).

27. Williams, W. H., *Delmarva's Chicken Industry: 75 years of progress* (Delmarva Chicken Association, 1998).
28. Bennett, C. E. et al., 'The broiler chicken as a signal of a human reconfigured biosphere', *Royal Society Open Science* (Dec 2018). https://doi.org/10.1098/rsos.180325
29. *The Chicken of Tomorrow Contest*, documentary (1948). www.youtube.com/watch?v=RrAV4zmZa-w
30. National Chicken Council, 'U.S. Broiler Performance 1925 to Present'. www.nationalchickencouncil.org/about-the-industry/statistics/u-s-broiler-performance/
31. Bennett, 'The broiler chicken as a signal of a human reconfigured biosphere'.
32. Ibid.
33. Ibid.
34. Baéza, E. et al., 'Influence of increasing slaughter age of chickens on meat quality, welfare, and technical and economic results', *Journal of Animal Science*, 90:6 (2012). https://doi.org/10.2527/jas.2011-4192

Chapter 8

1. Hunter, J. with notes by Bell, T., *Treatise on the natural history and diseases of the human teeth: explaining their structure, use, formation, growth, and diseases, in two parts* (Philadelphia: Haswell, Barrington, and Haswell, 1839), pp. 100–1.
2. 'History of Smallpox', Centers for Disease Control and Prevention: U.S. Department of Health & Human Services. www.cdc.gov/smallpox/history/history.html
3. Norkin, L. C., 'Ernest Goodpasture and the Egg in the Flu Vaccine' (2014). https://norkinvirology.wordpress.com/2014/11/26/ernest-goodpasture-and-the-egg-in-the-flu-vaccine/
4. Jacobson, T., 'Chickens and eggs: On the "Edison of Medicine" and his contribution to the study of vaccines', *The New Criterion* (1 April 2020). https://cet.org/wp-content/uploads/2020/04/New-Criterion-CET-Chickens-and-eggs-coronavirus.pdf
5. Hoyt, K., 'Vaccine Innovation: Lessons from World War II', *Journal of Public Health Policy*, 27:1 (2006), pp. 38–57. http://www.jstor.org/stable/3879064

6. Ibid.
7. Doshi, P., 'Trends in recorded influenza mortality: United States, 1900–2004', *American Journal of Public Health*, 99:8 (Aug 2009), pp. 1353–4. PMID: 18381993. www.ncbi.nlm.nih.gov/pmc/articles/PMC2374803/
8. Yeung, J., 'The US keeps millions of chickens in secret farms to make flu vaccines. But their eggs won't work for coronavirus', *CNN Health* (29 March 2020). https://edition.cnn.com/2020/03/27/health/chicken-egg-flu-vaccine-intl-hnk-scli/index.html
9. 'Changing dynamics in global poultry production', *Poultry World* (6 Jul 2010).
10. Gilbert, M., Slingenbergh, J. and Xiao, X., 'Climate change and avian influenza', *Revue scientifique et technique* (International Office of Epizootics), 27:2 (2008), pp. 459–66.
11. United States Government Accountability Office, 'AVIAN INFLUENZA: USDA Has Taken Actions to Reduce Risks but Needs a Plan to Evaluate Its Efforts', Report to Congressional Requesters (April 2017). www.gao.gov/assets/gao-17-360.pdf
12. 'Genetically modified chicken's eggs offer hope for cheaper drug production', *Pharmaceutical Technology* (28 Jan 2019). www.pharmaceutical-technology.com/news/genetically-modified-chicken-eggs/
13. https://en.wikipedia.org/wiki/Chick_culling
14. www.theguardian.com/commentisfree/2014/jan/17/would-you-prefer-to-eat-genetically-modified-eggs-or-see-day-old-chicks-destroyed
15. www.bbc.com/future/article/20171010-how-chicken-feathers-could-warm-our-homes
16. https://plantscience.agri.huji.ac.il/avigdor-cahaner/research-projects
17. Rotenberk, L., 'Flock Star: From Computers to Cars, Chicken Feathers Are Everywhere', *Modern Farmer* (28 Feb 2014). https://modernfarmer.com/2014/02/chicken-feathers/
18. Hazen, H. A., 'The Tornado: Appearances; Lieut. Finley's Views', *Science*, 15 (23 May 1890), p. 311. https://archive.org/details/jstor-1766014
19. Vonnegut, B., 'Chicken Plucking as Measure of Tornado Wind

Speed', *Weatherwise*, 28:5 (1975), p. 217. https://doi.org/10.1080/00431672.1975.9931768

20. Bacon, F., *The Works of Francis Bacon: Miscellaneous writings in philosophy, morality and religion*, Jones, M. (ed.) (London, 1815).
21. Aubrey, J., *'Brief Lives,' chiefly of Contemporaries, set down by John Aubrey, between the Years 1669 & 1696*, Vol. 1, Clark, A. (ed.) (Oxford: Clarendon Press, 1898). https://www.gutenberg.org/files/47787/47787-h/47787-h.htm
22. Bacon, *The Works of Francis Bacon*.
23. Thomas, L., 'The Shivering Chicken of Highgate', *The Evening News* (10 Dec 1957), p. 6.
24. 'OECD-FAO Agricultural Outlook 2021–2030'. www.oecd-ilibrary.org/agriculture-and-food/oecd-fao-agricultural-outlook-2021-2030_19428846-en
25. Ritchie, H. and Roser, M., 'Meat and Dairy Production', Our World in Data (2019). https://ourworldindata.org/meat-production
26. *World Agriculture: Towards 2015/2030* – An FAO perspective, The Food and Agriculture Organization of the United Nations (2002). www.fao.org/3/y4252e/y4252e05c.htm
27. www.bbc.co.uk/news/science-environment-28858289
28. https://sustainablefoodtrust.org/articles/we-need-to-talk-about-chicken-new-report-highlights-the-problems-with-intensive-chicken-production/
29. www.theguardian.com/society/2022/jan/20/antimicrobial-resistance-antibiotic-resistant-bacterial-infections-deaths-lancet-study
30. Chiandetti, C. and Vallortigara, G., 'Intuitive physical reasoning about occluded objects by inexperienced chicks', *Proceedings of the Royal Society B.*, 278:1718 (Jan 2011), pp. 2621–7. http://doi.org/10.1098/rspb.2010.2381
31. www.psychologytoday.com/gb/blog/smarter-you-think/202101/counting-chickens-right-after-they-hatch
32. Marino, L., 'Thinking chickens: a review of cognition, emotion, and behavior in the domestic chicken', *Animal Cognition*, 20 (2017), pp. 127–47. https://doi.org/10.1007/s10071-016-1064-4
33. Ghirlanda, S., Jansson, L. and Enquist, M., 'Chickens prefer beautiful humans', *Human Nature*, 13:3 (Sep 2002), pp. 383–9. https://doi.org/10.1007/s12110-002-1021-6

34. Marino, 'Thinking chickens'.
35. www.scientificamerican.com/article/the-startling-intelligence-of-the-common-chicken/

INDEX

D

E

V

W

Y

Z